I0868459

Casting Experiments and Microstructure of Archaeologically Relevant Bronzes

Quanyu Wang
Barbara S. Ottaway

BAR International Series 1331
2004

Published in 2016 by
BAR Publishing, Oxford

BAR International Series 1331

Casting Experiments and Microstructure of Archaeologically Relevant Bronzes

ISBN 978 1 84171 676 3

BAR Publishing is the trading name of British Archaeological Reports (Oxford) Ltd. British Archaeological Reports was first incorporated in 1974 to publish the BAR Series, International and British. In 1992 Hadrian Books Ltd became part of the BAR group. This volume was originally published by Archaeopress in conjunction with British Archaeological Reports (Oxford) Ltd / Hadrian Books Ltd, the Series principal publisher, in 2004. This present volume is published by BAR Publishing, 2016.

Printed in England

BAR
PUBLISHING

BAR titles are available from:

BAR Publishing
122 Banbury Rd, Oxford, OX2 7BP, UK
EMAIL info@barpublishing.com
PHONE +44 (0)1865 310431
FAX +44 (0)1865 316916
www.barpublishing.com

Contents

*The CD is no longer available. The contents are instead available for download from: http://www.barpublishing.com/additional-downloads.html

Figures

Tables

Plates

Acknowledgements

The project arose out of several years of teaching experimental casting and metal working to Masters and undergraduate students of the Department of Archaeology at the University of Sheffield. The students' enthusiasm was rewarding and stimulating but led to the realisation that more continuous work was needed to provide answers to questions on the behaviour of copper-based alloys used in prehistory. We would like to acknowledge the stimulus provided over the years by the students.

A substantial part of the work was conducted in the foundry of the Department of Engineering Materials at the University of Sheffield. We gratefully acknowledge the technical support of Phil Staton and Andy Marshall and the friendly help of the chief technician Ian Watts. Alan Harvey, an experienced foundry technician was enticed out of retirement for this project. He offered invaluable help, expertise and advice and made the project very enjoyable.

This work would not have been possible without the grant awarded to one of the authors (BSO) by the Leverhulme Trust which financed the post of a research associate, Dr Quanyu Wang, for two years from 1 December 2000 to 30 November 2002. This financial support is gratefully acknowledged.

The authors can be contacted by email as follows:
Quanyuwang@yahoo.com
B.ottaway@exeter.co.uk

1 Introduction

1.1 Aim of the study

Compositional data on bronze artefacts of the European Chalcolithic and Bronze Age are now quite numerous. Yet these data give no indication of *how* the bronzes were made: were they hammered into shape, for instance, or were they cast? If they were cast, in what kinds of mould were they cast? How quickly were the bronze artefacts cooled? Were they subjected to any cold or hot working after casting?

Modern metallurgical investigations are not usually interested in the copper-based alloys that were used by prehistoric smiths, because they are of little or no relevance to material used now. Most of the studies discussed in section 1.2 of alloys with compositions similar to the ones found in prehistory have been conducted using modern material for the moulds or adding fluxes, such as cuprex, and deoxidants to the melt. These oxidation and deoxidation techniques ensure gas-free, clean metal ready for pouring without degassing processes, but these modern materials were not available in prehistory. There is therefore scant information on the microstructure and behaviour of archaeologically relevant alloys cast with materials likely to have been accessible in the Chalcolthic and Early Bronze Age of Europe.

To build up a reference collection for the determination of ancient production methods of cast bronze artefacts, a series of casting experiments with archaeologically relevant alloys was carried out in the University of Sheffield. Bronze flat axes, characteristic of the European Chalcolithic to Bronze Age periods, were cast in moulds of sand, clay and bronze. The composition of the bronze, the moulding material and the cooling method after casting were systematically varied under controlled conditions. The microstructure, dendritic arm spacing or grain size and microhardness of the cast metals were studied on each casting. The malleability of the metals was also investigated by cold-rolling and annealing processes.

1.2 Review of previous work

Casting experiments have been carried out by metallurgists and archaeological scientists who wanted to explore the ancient production technology of bronzes (Bunk and Kuhnen 2003; Chadwick 1939; Faoláin and Northover 1998; Fasnacht 2001; Hanson and Marryat 1927; Lamm 1973; Matsuda 1928; Shalev 1999; Staniaszek and Northover 1983; Wirth 2003). However, systematic studies of castings with archaeologically relevant bronzes have rarely been conducted. Most of the published work focuses on one aspect of the casting alone, for instance on mould material, on cooling rate, or on the types of object cast in moulds.

Chadwick (1939) did casting experiments with bronzes high in tin, containing 14, 20 and 25% tin, and studied the effect of composition on the behaviour and on some of the physical properties of these bronzes. However, his casting experiments were carried out with phosphorus-containing deoxidant, which was not available in prehistory.

Northover and Staniaszek conducted a series of casting experiments using stone, clay and steel moulds in the early 1980s, but the results have not been fully published. Only bronzes with 10% tin and variable lead content (2, 5, 10, 15 and 20% Pb) have been published in detail (Staniaszek and Northover 1983).

Experimental work with arsenical copper has been carried out by Hanson and Marryat (1927) who studied solid solubility of arsenic in copper. Budd (1991) experimentally cast copper containing between 0.5 and 12 % arsenic in open sand and steel moulds, subjected the resulting alloys to cold working and heat treatment and subsequentially studied the microstructure and mechanical properties and their implication for manufacturing techniques (Budd and Ottaway 1991, 1995). Northover (1989), Lechtman (1996) and Lechtman and Klein (1999) followed similar paths in the quest to unravel the production, properties and behaviour of arsenical copper.

More recently Junk (2003) and Schmalfuss and Pernicka (in press) studied properties and behaviour of copper with the characteristic *fahlerz* elemental composition, i. e. copper containing arsenic, antimony and bismuth, in experimentally produced and on prehistoric material.

Some of the published data on experimental casting with archaeologically relevant copper alloys , specifically tin and tin-lead bronzes, are listed in Table 1.1.

Experimental and analytical methodology will be discussed in chapter 2. This is followed by three chapters describing the casting and metallography of bronzes cast in sand, clay and bronze moulds (chapters 3, 4 and 5 respectively). The microstructures of as-cast bronzes from the three different moulding materials are compared in chapter 6 and their malleability and different cooling rates are discussed in chapter 7. Chapter 8 examines the behaviour of tin and lead during melting and casting. Finally, in chapter 9, the results are summarised and questions that have been raised by the project and that need further research are pointed out.

All figures and a selection of photomicrographs are printed in black and white in this volume. They are reproduced in colour in appendix 1 on the accompanying CD. There is one plate of colour photomicrographs in chapter 9. The complete series of colour photomicrographs for bronzes cast in sand moulds is given in appendix 2, that for bronzes cast in clay moulds is in appendix 3 and that for bronzes cast in bronze moulds is in appendix 4. The photomicrographs of the series of cold-working and annealing are given in appendix 5. All sample numbers discussed in this volume, with their composition and cooling regimes, are given in appendix 6.

Table 1.1 ***Published data on experimental casting of bronzes***

Composition (wt %)		Technique	Description	References
Sn	Pb			
10	–		Experimental casting using metal, stone and clay moulds	Staniaszek and Northover 1983
10	2		Experimental casting using metal, stone and clay moulds	Staniaszek and Northover 1983
10	5		Experimental casting using metal, stone and clay moulds	Staniaszek and Northover 1983
10	10		Experimental casting using metal, stone and clay moulds	Staniaszek and Northover 1983
10	15		Experimental casting using metal, stone and clay moulds	Staniaszek and Northover 1983
10	20		Experimental casting using metal, stone and clay moulds	Staniaszek and Northover 1983
21	–	Quenched	Modern bronze from Philippines	Northover 1996: fig. 4a
22	–	Quenched	Modern bronze from Philippines	Northover 1996: fig. 4b
23	–	Quenched	Modern bronze from Philippines	Northover 1996: fig. 4d-h
24	–	Quenched	Modern bronze from Philippines	Northover 1996: fig. 4c
22	6	As-cast	Experimental casting	Scott 1991: fig. 45
24	–	Quenched at 650°C	Experimental casting	Scott 1991: fig. 46
–	20		Experimental casting	Northover 1996: fig. 12

2 Experimental and analytical methodology

2.1 Experimental strategies

The compositional analyses of 22,000 early European bronzes were published in the SAM volumes in the 1960s and 1970s (Junghans *et al.* 1960, 1968 and 1974). The majority of these bronzes contain less than 10% tin and leaded bronzes are rare.

In the light of these data, and to avoid repeating other researchers' work, 12 different alloys (tin bronzes and leaded tin bronzes) were chosen for the experiments discussed here to improve our understanding of archaeological bronzes. The proposed compositions of the selected alloys are listed in table 2.1.

The quality of castings, reflected by the microstructure, is affected by the size of the cast object. The larger the object, the longer it takes to solidify and the more likely it is to be porous. Conversely, porosity decreases with fast solidification. This can be demonstrated by comparing the microstructures of an experimentally cast bronze plate of 10 x 5 x 0.5 mm and an experimentally cast palstave. The former is less porous than the latter.

For the experimental series presented here, a flat axe, typical of the European Chalcolithic to Early Bronze Age, was used for all castings to obtain standardisation and comparability of results. For the production of all three types of mould a pattern was made of pine wood and painted with shrink-resistant varnish (fig. 2.1).

Two different cooling methods, air-cooled and water-quenched, were used in the sand-mould casting series. For clay moulds, three different ways of treating the moulds prior to and after the molten metal was poured in produced three different cooling rates: one was preheated (to 350°C) and air-cooled; one was not preheated and was air-cooled; the third was not preheated and was water-quenched. For casting experiments using bronze moulds, water-quenched casting was not possible because of safety regulations. Two cooling methods were thus used for bronze-mould casting: one in a preheated mould (to 350°C) and one in a flame-warmed (not preheated) mould. The cooling methods for the casting series are listed in table 2.1.

Table 2.1 ***Metal compositions and cooling methods of the experimental casting series in sand moulds***

Mould material	Composition (wt %)			Cooling methods		
	Cu	Sn	Pb	Preheated to 350°C /air-cooled	Unpreheated /air cooled	Unpreheated /water-quenched
Sand	98	2	-		v	v
	94	6	-		v	v
	90	10	-		v	v
	85	15	-		v	v
	77	23	-		v	v
	96	2	2		v	v
	92	2	6		v	v
	88	2	10		v	v
	92	6	2		v	v
	88	6	6		v	v
	84	6	10		v	v
	88	10	2		v	v
	98				v	v
Clay	98	2	-	v	v	v
	94	6	-	v	v	v
	90	10	-	v	v	v
	85	15	-	v	v	v
	77	23	-	v	v	v
	96	2	2	v	v	v
	92	2	6	v	v	v
	88	2	10	v	v	v

Table 2.1 (contd.) ***Metal compositions and cooling methods of the experimental casting series in sand moulds***

	92	6	2	v	v	v
	88	6	6	v	v	v
	84	6	10	v	v	v
	88	10	2	v	v	v
	98			v	v	v
Bronze	98	2	-	v	v	
	94	6	-	v	v	
	90	10	-	v	v	
	85	15	-	v	v	
	77	23	-	v	v	
	96	2	2	v	v	
	92	2	6	v	v	
	88	2	10	v	v	
	92	6	2	v	v	
	88	6	6	v	v	
	84	6	10	v	v	
	88	10	2	v	v	
	98			v	v	

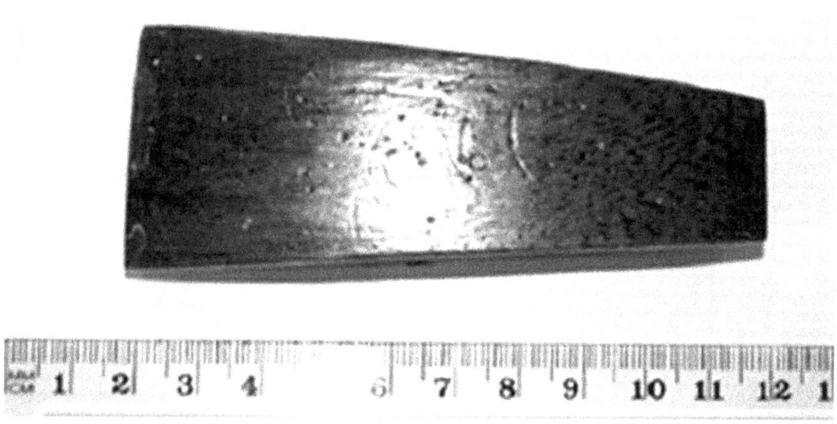

a. Face

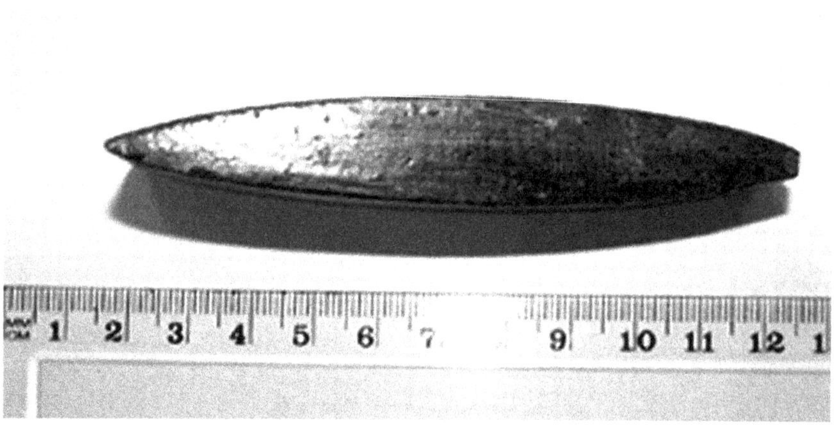

b. Side

Figure 2.1 The wooden pattern used in the experiments

2.2 Casting

There are many factors involved in the casting of a metal object, many of which were tested in initial experiments to produce a high-quality casting. The parameters tested included type of furnace, mould size, compactness of sand moulds, number of vents in the mould, size of feeder, casting temperature, size of charcoal pieces, additions to the casting, cooling rates of the casting, and sampling location. The variables common to all three types of mould are described here while those variables that are specific to each individual type of mould will be discussed later in the relevant chapters.

A plumbago (clay graphite) crucible was used for melting the metal. The chemical composition of a plumbago crucible is: 37-43% C, 9-16% SiC, 12-17% Al_2O_3. The crucibles used in our experiments were 165 mm high; the diameter at the top was 130 mm and at the bottom 95 mm. Copper ingots, approximately 10-30 x 10-30 x 2 mm, and almost spherical copper slugs with a diameter of *c.* 10 mm and 5 mm thick, were tested. The capacity of the crucible was given by the manufacturer as 8 kg. However, the actual capacity depends on the size of the raw material. Only 3-4 kg of the copper ingots but 5-6 kg of the copper slugs could be fitted into a crucible for a single charge.

Two types of furnace, a gas furnace and an induction furnace, were used for the initial experiments. The gas furnace that was available was found to be too slow. A charge of 3 kg took 3.5 hours to reach a temperature of 1050°C, which is below the melting point of copper. An induction furnace was chosen for the experiments so that the metal heated more quickly, thereby also reducing the amount of contamination entering the melt. Copper can be heated in such a furnace to a temperature of 1200-1300°C in 20 minutes.

Casting temperature is the temperature of the molten metal just before it is poured into the mould. However, it is difficult to measure the precise pouring temperature, since the molten metal needs to be poured seconds after it has been taken from the furnace. In this volume *melting temperature* refers to the temperature of molten copper in the furnace just before the furnace is turned off. The temperature of the molten metal drops dramatically during the addition of alloying components, removal of the crucible, and the skimming off of the charcoal residue. For example, molten copper dropped from 1280°C to 1180°C in just one minute. The question is: how high should the casting temperature be? Campbell (1991, 4) has reported that, for an aluminium alloy (T_{mp} = 660°C), casting can be done successfully at a temperature of 700-750°C, which is about 100°C higher than the melting temperature of aluminium. Experiments with bronze carried out by Ratka (1996, 15) suggest that a casting temperature of 1150°C is necessary for a flawless casting into an unpreheated mould and our initial experiments have also indicated that a high *casting temperature* produces a better quality casting. To keep all casting parameters as constant as possible, we kept the *melting temperatures* of the copper in the range of 1270-1290°C. Alloying components such as tin and lead were added just before the crucible was lifted out of the furnace. The *casting temperatures* were in the range of 1170-1190°C for all experiments.

In modern casting procedures, cuprex (a flux containing mainly manganese ore, fluorspar, silica sand and sodium carbonate) and a deoxidant (15% P-Cu) are used to produce sound castings of copper-based alloys. Cuprex is added with the charge to expel hydrogen. Deoxidant is added to the molten copper before pouring to prevent oxidation of the metal. We tried to avoid using these additions since we aimed to simulate prehistoric casting as closely as possible and granulated charcoal was therefore added at both the bottom and the top of the charge to serve as a reducing agent. The size of the charcoal was in the range of 5-10 mm. For an average charge of 3 kg copper in the crucible 60-100 g charcoal was found to be adequate. The results of the experiments show that charcoal is satisfactory as a deoxidant. The cast artefacts were almost free of inclusions but porosity continued to be a problem, which could, however, be greatly reduced by a combination of controlling the mould construction, e.g. number of vents, feeder size, and casting temperature and cooling rates.

Induction furnaces work on the principle of electromagnetic induction. The charge is heated by the secondary currents induced by a primary coil placed in the furnace (Davies and Oelmann 1985, 74). Turbulence of the charge during melting is one of the major influences on casting quality. This was confirmed by our initial experiments, as can be seen in figures 2.2 and 2.3, which show the microstructure of bronzes with 6% tin cast in sand moulds. The first cast (fig. 2.2) is the result of pouring the molten metal into the mould as soon as the furnace was switched off, whereas the second cast (fig. 2.3) was allowed to rest for 30 seconds before being poured. A comparison of the two microstructures shows that allowing the metal to rest produces a much better cast. The crucible with the charge should therefore be kept in the furnace for a short time after the furnace is turned off to allow gases such as oxygen, hydrogen, carbon dioxide and carbon monoxide, etc. to escape before it is lifted out for pouring. To prevent the temperature of the molten metal from dropping too low, the resting time of the metal was set to 30 seconds.

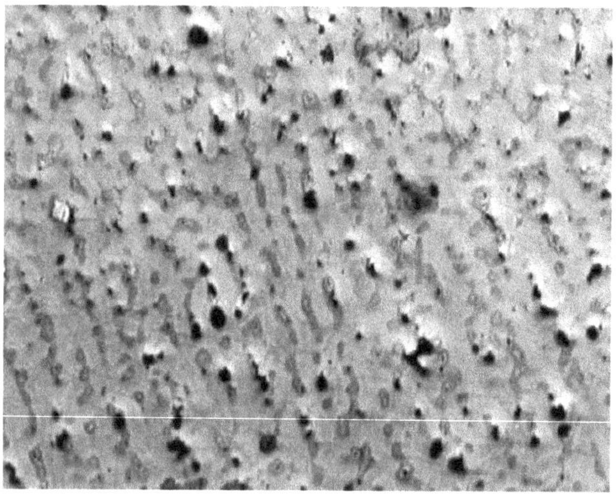

Figure 2.2 Photomicrograph of cast object from initial experiment, showing extensive porosity: 6% Sn, air-cooled (image width 1.3 mm)

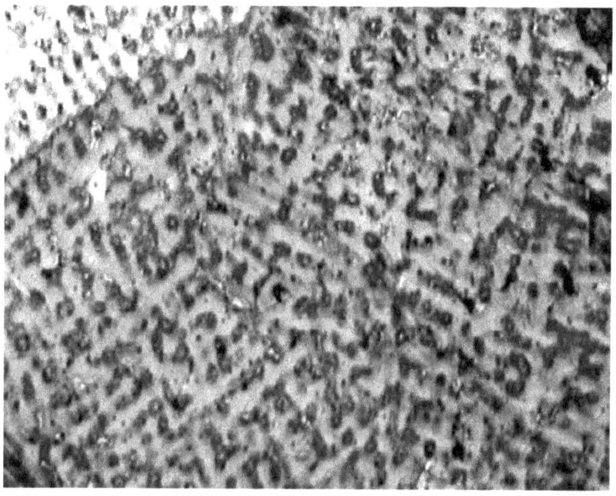

Figure 2.3 Photomicrograph of 6% Sn bronze, air-cooled, showing little porosity (image width 1.3 mm)

For a valid comparison of the experimental results, the casting conditions must be kept constant. This was achieved by the following methods:

Melting procedure

- Charcoal was added both at the bottom and at the top of the crucible to prevent oxidation of the copper. The amount of the charcoal was between 60-100 g depending on the charge (fig. 2.4).
- The copper was heated in the induction furnace to a temperature of 1250-1280°C in about 20 minutes. The crucible was covered with a lid during heating (fig. 2.5).
- A thermocouple of NiAl/NiCr was used to measure the temperature of the metals (fig. 2.6). The crucible was left in the furnace to rest for 30 seconds after the furnace was turned off to reduce the gas content.
- Tin and lead were added just before the crucible was lifted out for pouring to avoid oxidation.

Casting

- The molten metal was stirred and the charcoal residue was skimmed off before it was poured (fig. 2.7).
- The metal was poured until it filled the mould to the top of the feeder. The pouring sequence was: preheated mould, air-cooled mould and finally the one that was quenched in water.
  For the casts that were to be water-quenched, the moulds were broken open and the objects were immersed in cold water within two minutes of pouring.

Reduction in thickness of the metals was achieved by a cycle of cold-rolling and annealing. The procedures will be described in detail in chapter 7.

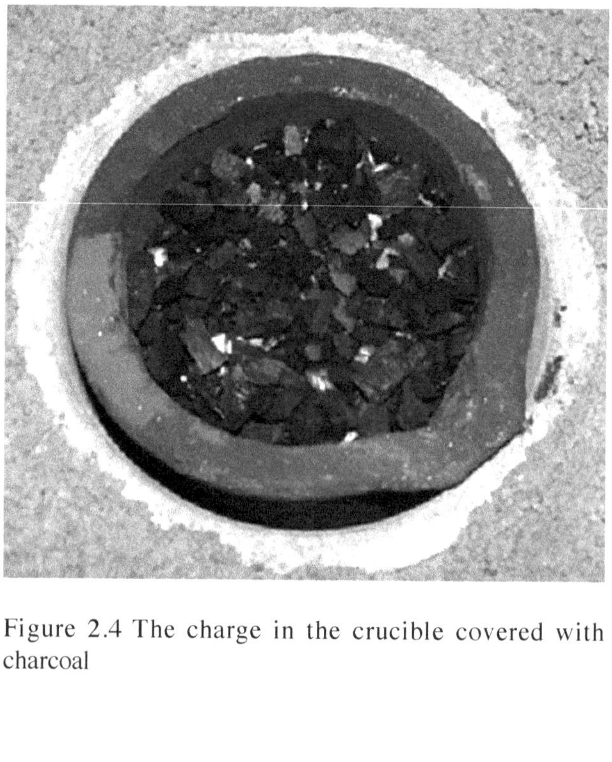

Figure 2.4 The charge in the crucible covered with charcoal

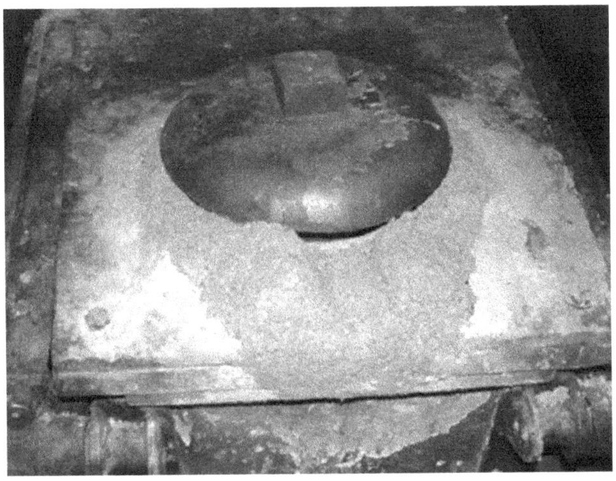

Figure 2.5 A lid covered the crucible during melting

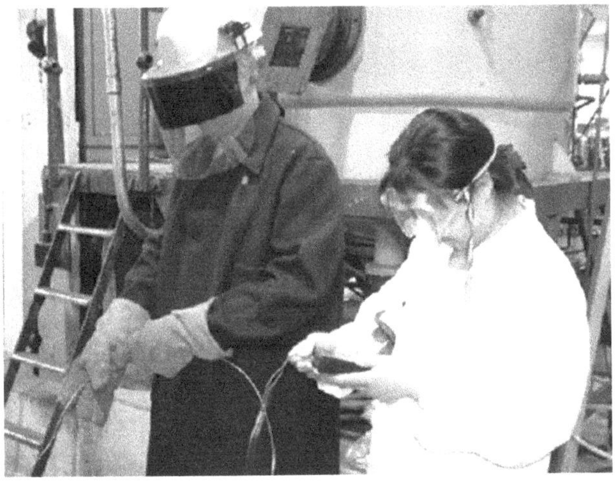

Figure 2.6 Temperature of the melts measured with a thermocouple

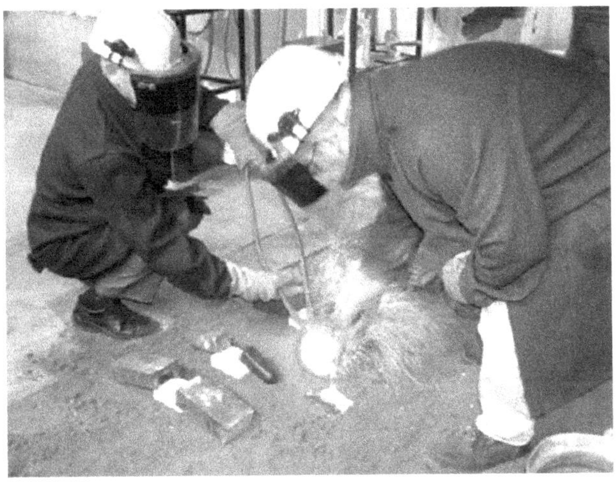

Figure 2.7 Charcoal residue and dross skimmed off before pouring

2.3 Sampling and analyses

All cast axes were sectioned across the middle for metallography. The metallographic samples were prepared by the conventional method, i.e. polished to 0.25 mm finish and etched with alcoholic ferric chloride. Photomicrographs were taken using an Olympus microscope equipped with an image-grabbing system. Since photomicrographs are much more informative in colour than in black and white, the colour versions of the black and white figures given in the text are included on the accompanying CD.

Compositional analyses were conducted by ICP at the Sheffield Analytical Services. A small number of microanalyses were also carried out using a JEOL JSM-6400 Scanning Electron Microscope equipped with EDS for confirmation of the metallographic observations. The operating condition was 20KV. The microanalyses show the distribution of alloying elements in different phases; they do not refer to the average composition of the cast object.

The grain size of granular structures and the dendrite arm spacing (DAS) of dendritic structures were measured by the intercept method (Scott 1991, 51). Here dendrite arm spacing refers to secondary dendrite arm spacing, because it is the most important structural length measurement. The mechanical properties of most alloys are strongly dependent on secondary arm spacing: strength, ductility and elongation increase as DAS decreases (Campbell 1991, 146).

The microhardness was measured using a Leco M-400 hardness tester, taking the average of five readings across the metallographic section from one surface to the other at equal intervals. The microhardness values refer to the Vickers Scale (H_V).

The results of the casting experiments and of the metallographic analyses are presented in the following three chapters.

3 Casting and metallography of bronzes cast in sand moulds

3.1 Introduction

Numerous bronze artefacts were cast in the Bronze Age, yet the number of extant moulds in which they could have been produced does not match the number of the Bronze Age artefacts found. It has been proposed (Ottaway and Seibel 1997, 59) that, apart from stone and clay moulds, which have survived in the archaeological record, sand moulds, which would have disintegrated completely after use, were also used. Casting experiments using sand moulds have been conducted for some time at the University of Sheffield and a few papers on this topic have been published (Ottaway and Seibel 1997, 59; Eccleston and Ottaway 2002). This present work is a systematic study of microstructure and of the effect of cold-working on bronzes cast under controlled conditions.

3.2 Mould-making

There are several factors in the production of a sand mould which greatly influence the quality of the object cast in the mould. They are:

1 clay content of the sand
2 water content of the sand
3 compactness of the sand
4 size of feeder and runner bush
5 size and position of the vents in the mould.

They will be discussed below because, although they are well known to experienced casting and foundry workers, they are not often discussed in the archaeological literature.

1. The quality and clay content of the sand are two of the important factors in producing casts of good quality. To have bonding property, the sand has to contain 3-6% clay. Too much clay reduces the permeability of the mould, trapping gases in the metal and leading to a porous cast. Initial experiments have shown that using facing sand with a high clay content, for example pure Mansfield sand, which contains approximately 12% clay, results in a very porous cast. In the series of experiments reported here, floor sand with the addition of new Mansfield sand was used for making the moulds.

2. The moisture content of the moulding sand also influences the quality of the cast (Eccleston and Ottaway 2002). Theoretically, the moisture content should be about 8%. In practice, water was added to the sand gradually until it held together when squeezed in the hand. Too much moisture reduces the permeability of the sand, increases the production of steam, and is dangerous, as it can cause the mould to explode. The moisture of the sand should therefore be kept as low as possible to assure a high-quality cast.

3. The compactness of the sand affects the quality of the casting. Ramming the sand into the box too hard reduces its permeability and results in porous artefacts. On the other hand, ramming it too little increases the risk of the mould collapsing and increases the surface roughness of the artefact as well. It was evident from the initial experiments that a fairly low compactness produced better cast objects. The sand was therefore rammed relatively lightly. The density of the mould was measured with a Ridsdale B Scale Green Hardness Tester. It was found that a mould with a reading of 50 units on the 'B' Scale produced good-quality casting. (It should be noted that these readings depend on how hard the tester is pressed on to the surface of the sand mould: they could be different if they are measured by different persons. They are thus only useful for standardisation if there has been a single operator.)

Two sizes of sand mould were tested initially to investigate the influence of mould size on the quality of the cast. The dimension of the small one was 17 x 17 x 10 cm, and that of the large one 25 x 25 x 10 cm. The cavity for the casting was made in the centre of the box, the depth of which was the same in both cases. Probably because of the small size of the object in comparison to the mould box, the castings show little difference. The larger size of box was chosen for all the rest of the experiments.

4. A feeder was made for the metal to flow into the cavity and to provide a reservoir of molten metal to compensate for contraction on solidification. Filling is different from feeding; the former refers to filling the cavity in the mould, the latter to making good the loss through shrinkage during solidification. Filling normally takes seconds, whereas feeding takes minutes (Campbell 1991, 179). The optimum feeder size is difficult to calculate accurately (Campbell 1991, 180). Two sizes of feeder were tested in the initial experiment, a smaller cylinder tube with a diameter of 16 mm and a larger one with a diameter of 21 mm. The results showed that the cast made with the large feeder was of better quality.

The running system, i.e. the passage through which the molten metal flows into the mould, is one of the most important factors determining the quality of the cast. Top or side gates, which allow the metal to fall freely and splash into the mould cavity, should be avoided (Campbell 1991,

33). In the experiments discussed here, a gate at one side of the cavity was created which allowed the metal to fill the mould from the side. A runner bush was fitted slightly offset to prevent the metal from falling freely into the cavity of the mould.

5. Vents were made to release gases from the mould. At first two vents were cut at the wide end of the cavity. This produced porous casts. Increasing the number of vents to a total of nine, three at each end and three on the side away from the feeder, and perforating the sand to create three small holes in the top of the mould improved the surface of the cast considerably.

With these five major points in mind, the moulds for the series of casting experiments in sand were constructed according to the following principles:

- adding as little Mansfield sand as possible to the floor sand to keep the clay content of the moulding sand low;
- keeping the sand as dry as possible while adding sufficient water to enable the mould material to hold its shape;
- compressing the sand as lightly as possible to retain the permeability of the mould;
- making a good feeder and an offset runner bush;
- making a total of nine vents on three sides of the cavity, and adding three small holes in the top of the mould to allow gases to escape during pouring and solidification.

In the light of all these points, the sand moulds for the series of experiments were produced as follows:

- Modern foundry two-piece moulding boxes were used in the experiments.
- The drag box was positioned on the working table with the varnished wooden pattern suitably placed and dusted with parting powder (fig. 3.1).
- The pattern was covered with sieved facing sand – a mixture of used sand and about 20% new Mansfield sand. The box was then filled with moulding sand: floor sand, i.e. used sand, mixed with a little new Mansfield sand.
- The sand was rammed and more sand was added and rammed until the box was completely filled. The sand was compressed by a flat ram and strickled flat.
- The drag box was turned over, and the cope box placed on it, the surface dusted with parting powder and a feeder suitably placed (fig. 3.2).
- The same filling and ramming procedure as before was used to fill the cope box.
- The feeder was removed and three small holes were made (fig. 3.3).
- The cope box was lifted off and the parting powder was brushed off. The pattern was taken out by wetting around its edge and lightly tapping it. A runner and vents were carefully made in the sand (fig. 3.4). The mould was closed and a runner bush fitted (fig. 3.5).

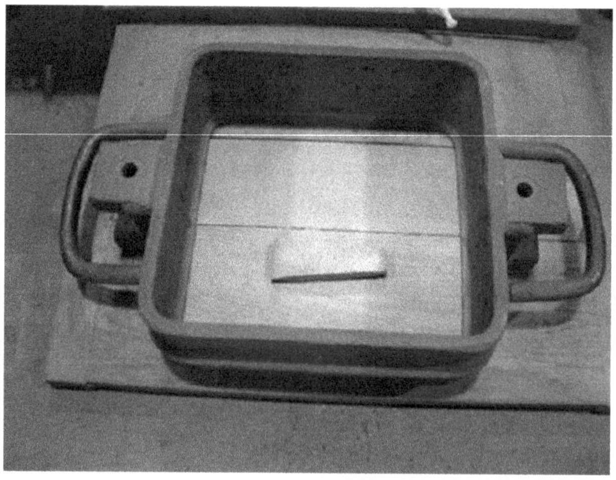

Figure 3.1 The pattern placed in the drag box

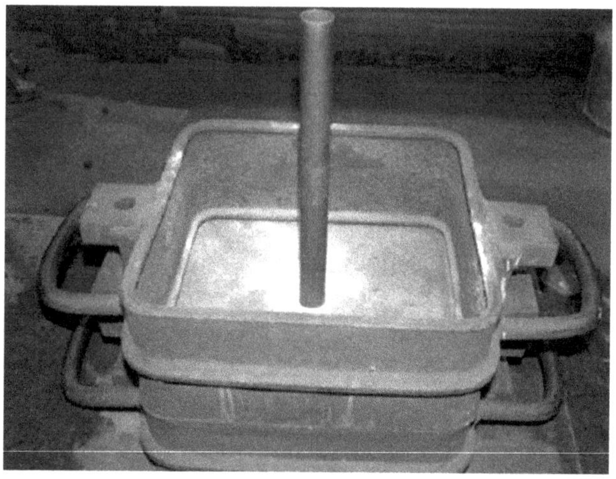

Figure 3.2 The cope box placed on the drag box and the feeder put in place

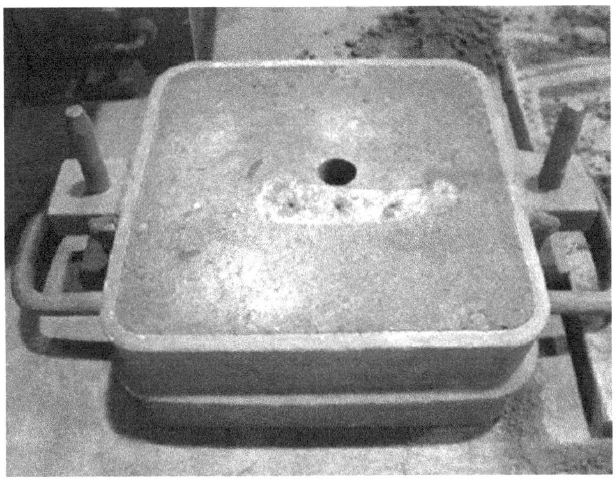

Figure 3.3 Both drag and cope box filled with sand, three holes made and the feeder removed

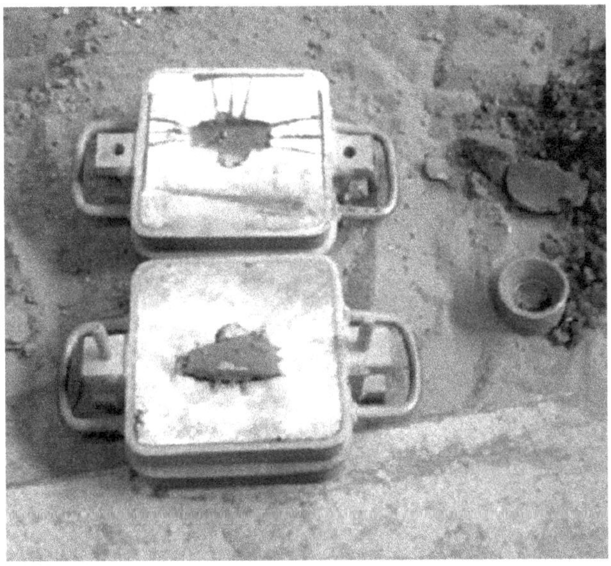

Figure 3.4 Mould opened after cooling, showing position of the nine vents and ingate

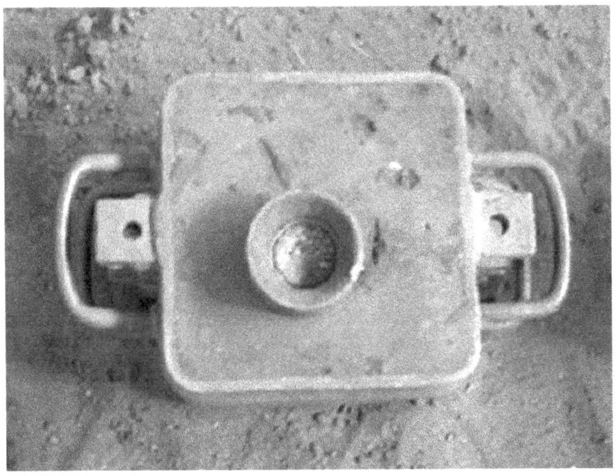

Figure 3.5 Metal filled up to the runner bush

3.3 Experimental casting

In the initial experiments, three different cooling rates were used for the sand mould casting, which included cooling in embedded sand, in air, and in water. As the casts embedded in sand were found to be very porous, this cooling method was discarded in the casting series.

Melting the metal and controlling the temperature were conducted as described in section 2.2. The results of the casting experiments in sand moulds are presented here.

3.4 Results and discussion

In general, there was very little flashing except at the positions where the vents had been placed (figs. 3.4 and 3.6). The cast axes had sharply defined edges and their size was almost the same as that of the wooden pattern (fig. 3.7).

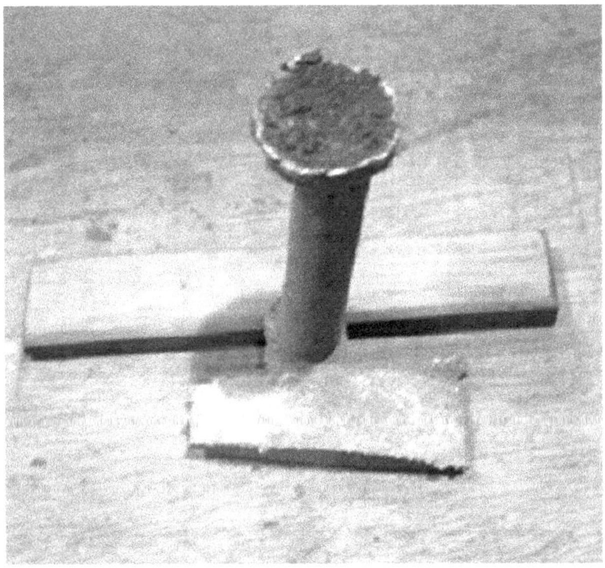

Figure 3.6 A cast axe with the feeder still attached

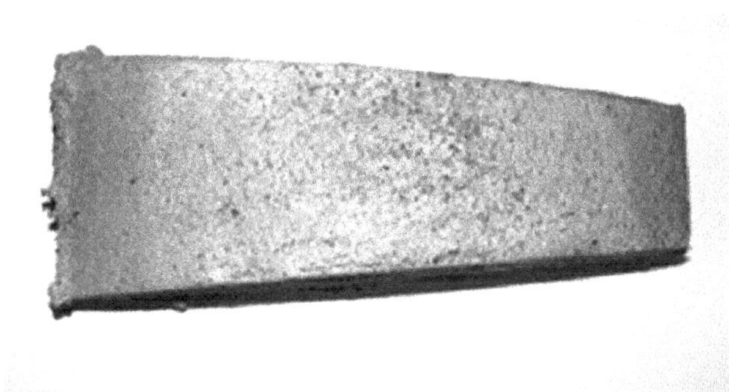

Figure 3.7 A finished axe after the feeder was removed and the axe sand-blasted

3.4.1 Microstructure

Micropores were found to be present in most of the sections. This agrees with simulated and experimental casts reported by other researchers (Ratka 1996, 302; Staniaszek and Northover 1983; Swiss and Ottaway in press).

The microstructures of the cast bronzes in the present study are complex. A selection of black and white photomicrographs are shown in figures 3.8-3.21. They can also be seen in colour in appendix 1 on the accompanying CD. A complete series of photomicrographs of bronzes cast in sand moulds can be found in appendix 2 on the CD. The results are summarised in tabular form in table 3.1. The details are presented and discussed below.

Table 3.1 ***Composition, cooling regimes and microstructures of bronze axes cast in the sand moulds***

Sample no.	Composition[1] (wt %)		Cooling method	Microstructure	Figure no.
	Sn	Pb			
41	2	—	Water-quenched	Granular without α+δ eutectoids	3.80
42	2	—	Air-cooled	Granular without α+δ eutectoids	3.90
35	6	—	Water-quenched	Dendritic with α+δ eutectoids	3.11
36	6	—	Air-cooled	Dendritic with very few α+δ eutectoids	3.12
43	10	—	Water-quenched	Dendritic with α+δ eutectoids; δ is dark.	3.13
44	10	—	Air-cooled	Dendritic with α+δ eutectoids	3.14
37	15	—	Water-quenched	Dendritic with many α+δ eutectoids; δ on boundaries and needle b within grains.	3.16
38	15	—	Air-cooled	Dendritic with many α+δ eutectoids	3.15
39	23	—	Water-quenched	Granular, dark δ on boundaries and needle β within grains	3.18 and 3.19
40	23	—	Air-cooled	Many α+δ eutectoids	3.17
45	2	2	Water-quenched	Both dendritic and granular; no α+δ eutectoids	3.21
46	2	2	Air-cooled	Both dendritic and granular; no α+δ eutectoids	—
47	2	6	Water-quenched	Both dendritic and granular; no α+δ eutectoids	—
48	2	6	Air-cooled	Both dendritic and granular; no α+δ eutectoids	3.10
49	2	10	Water-quenched	Granular, no α+δ eutectoids	3.22
50	2	10	Air-cooled	Granular, no α+δ eutectoids	—
51	6	2	Water-quenched	Dendritic with α+δ eutectoids; dark δ	—
52	6	2	Air-cooled	Dendritic without α+δ eutectoids	—
53	6	6	Water-quenched	Dendritic with a few α+δ eutectoids	—
54	6	6	Air-cooled	Dendritic without α+δ eutectoids	—
55	6	10	Water-quenched	Dendritic without α+δ eutectoids	—
56	6	10	Air-cooled	Dendritic without α+δ eutectoids	—
57	10	2	Water-quenched	Dendritic with many α+δ eutectoids; dark δ	—
58	10	2	Air-cooled	Dendritic with some α+δ eutectoids	—

Note: [1] The balance is copper.

Metallographic examination has shown that all cast bronzes with a low tin (2% Sn) content, including the leaded bronzes, have a granular structure (figs. 3.8-3.10) and the cast bronzes with 6, 10 and 15% tin, including the leaded tin bronzes, have a dendritic structure, as illustrated in figures 3.11 and 3.12. In principle, the microstructure of casts of known composition can be predicted on the basis of phase diagrams. However, phase diagrams refer to an equilibrium situation in which the metals cool very slowly. In the experimental casting and, as has been suggested by Scott (1991, 11), in metals cast in prehistory, casts were cooled rather faster than equilibrium solidification.

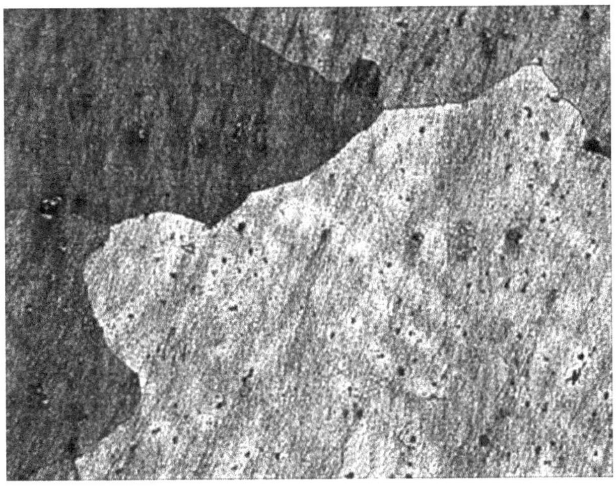

Figure 3.8 Photomicrograph of 2% Sn bronze; water-quenched, showing a granular structure with coring (image width 0.65 mm)

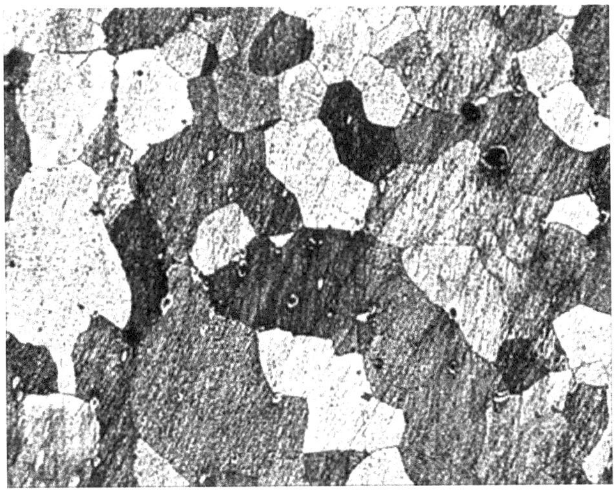

Figure 3.9 Photomicrograph of 2% Sn bronze; air-cooled, showing a granular structure without obvious coring (image width 0.65 mm)

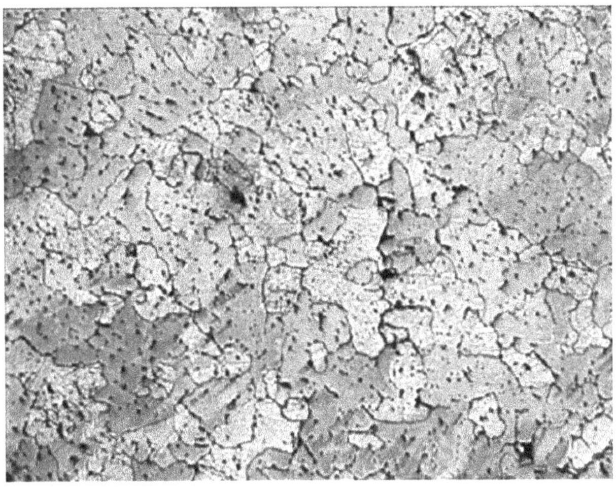

Figure 3.10 Photomicrograph of 2% Sn + 6% Pb bronze; air-cooled, showing a granular structure and pores on the grain boundaries (image width 1.3 mm)

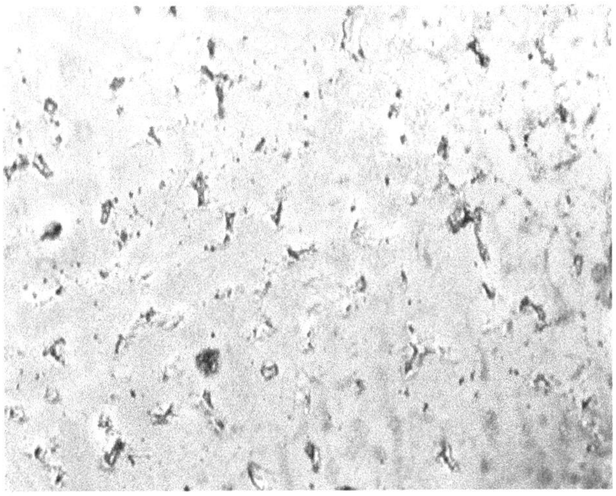

Figure 3.11 Photomicrograph of 6% Sn bronze; water-quenched, showing a dendritic structure with $\alpha+\delta$ eutectoids (image width 0.65 mm)

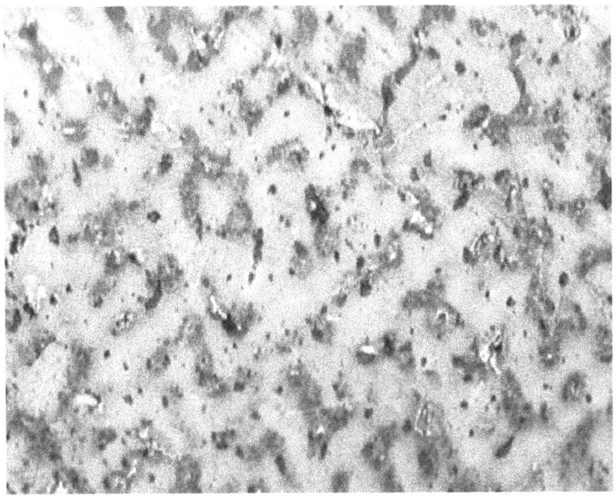

Figure 3.12 Photomicrograph of 6% Sn bronze; air-cooled, showing a dendritic structure with very few $\alpha+\delta$ eutectoids (image width 0.65 mm)

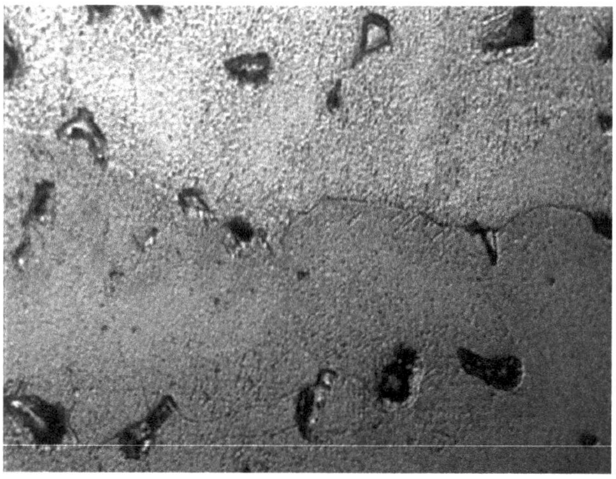

Figure 3.13 Photomicrograph of 10% Sn bronze ; water-quenched, showing dark α+δ eutectoids (image width 0.13 mm)

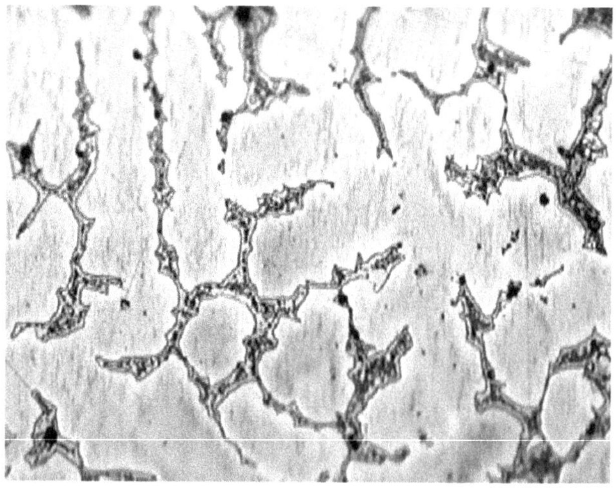

Figure 3.15 Photomicrograph of 15% Sn bronze; air-cooled, showing many α+δ eutectoids in the interdendritic regions (image width 0.33 mm)

The microstructure of the water-quenched bronzes is very different from that of the air-cooled ones. For example, the water-quenched 2% tin bronze has evident coring (fig. 3.8) which was not seen in the air-cooled one (fig. 3.9). In the 6% tin bronzes, the water-quenched one has more α+δ eutectoids (fig. 3.11) than the air-cooled sample of the same composition (fig. 3.12). The δ phase in water-quenched casts looks dark under the microscope (fig. 3.13) compared with the light blue of the air-cooled sample (fig. 3.14). The air-cooled 15% tin bronze has many α+δ eutectoids (fig. 3.15), while the water-quenched 15% tin bronze has a β phase both in the interdendritic regions and within the α grains (fig. 3.16).

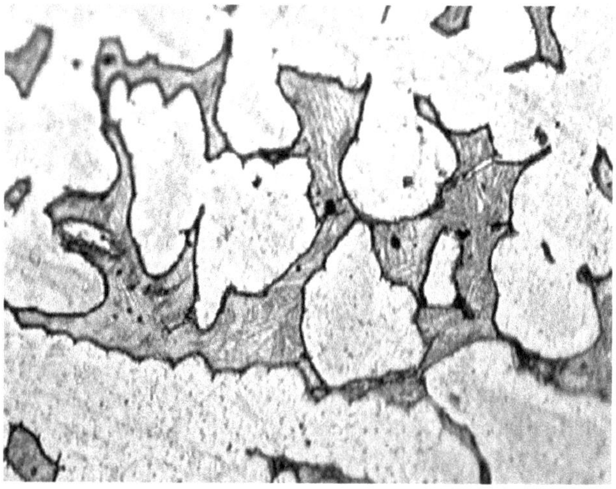

Figure 3.16 Photomicrograph of 15% Sn bronze; water-quenched, showing needle β in the interdendritic regions; the dark phase surrounding the needle β is likely to be δ phase (image width 0.33 mm)

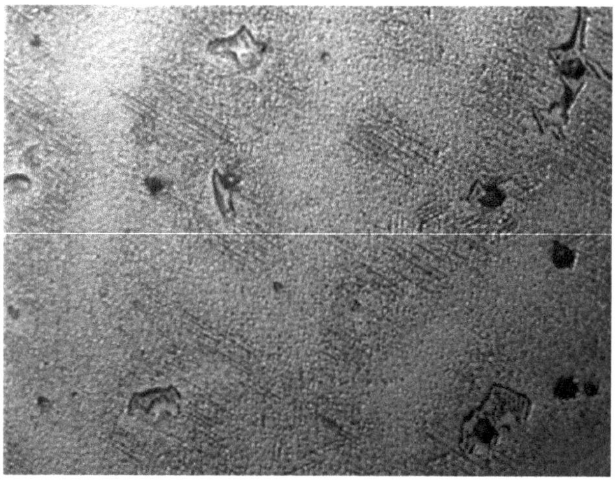

Figure 3.14 Photomicrograph of 10% Sn bronze; air-cooled, showing islands of α+δ eutectoids (image width 0.13 mm)

Bronze containing 23% tin is quite different from the other tin bronzes discussed here. In the air-cooled bronze $\alpha+\delta$ eutectoids are dominant (fig. 3.17). The water-quenched bronze has a granular structure, consisting of β needles within the grains and dark phase on the grain boundaries (figs. 3.18 and 3.19). The dark phase on the grain boundaries is likely to be δ phase. The temperature at which the bronzes were quenched was not measured in this experiment but may be estimated from the phase diagram and published data. The microstructures of bronzes (fig. 3.20) with similar levels of tin, quenched at different temperatures, have been recorded by Matsuda (1928). In these figures (fig. 3.20a-c) one can see the difference in microstructure that results from different quenching temperatures for bronzes with a constant tin content. The precise tin content of the bronze shown in figures 3.18 and 3.19 is 23.09% (see table 8.1, batch 40). The Cu-Sn phase diagram (Scott 1991, 122), with its lack of an α phase, suggests that the bronze was probably quenched at 650-750°C.

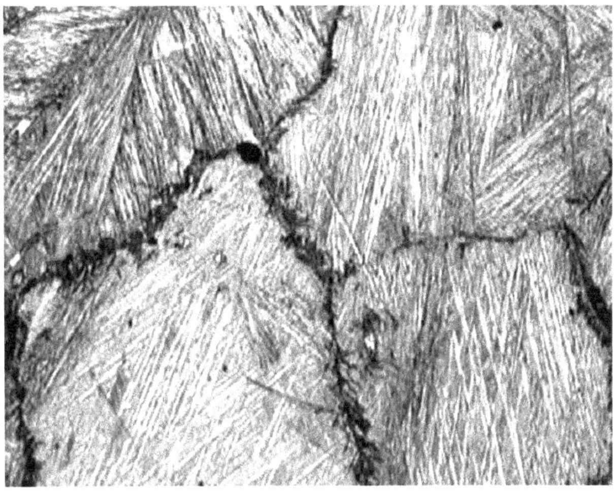

Figure 3.18 Photomicrograph of 23% Sn bronze ; water-quenched, showing dark grain boundaries and needle β within the grains (image width 1.3 mm)

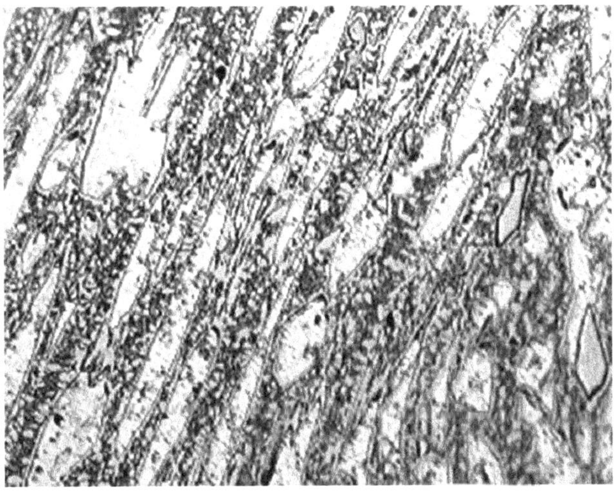

Figure 3.17 Photomicrograph of 23% Sn bronze; air-cooled, showing very many $\alpha+\delta$ eutectoids (image width 0.33 mm)

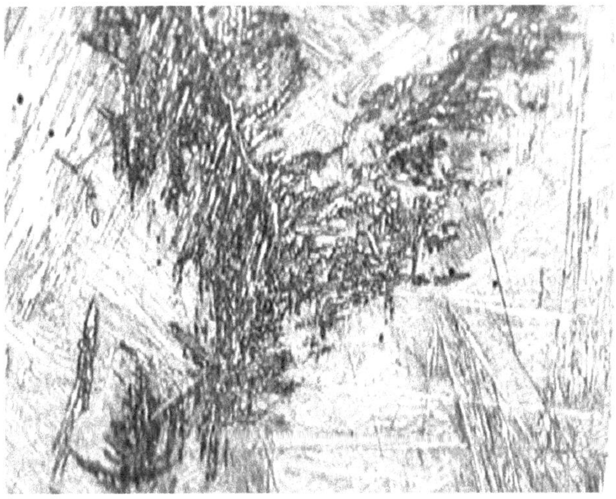

Figure 3.19 Photomicrograph of the same sample as in figure 3.18 at higher magnification. The dark phase on the boundaries is likely to be δ phase (image width 0.33 mm)

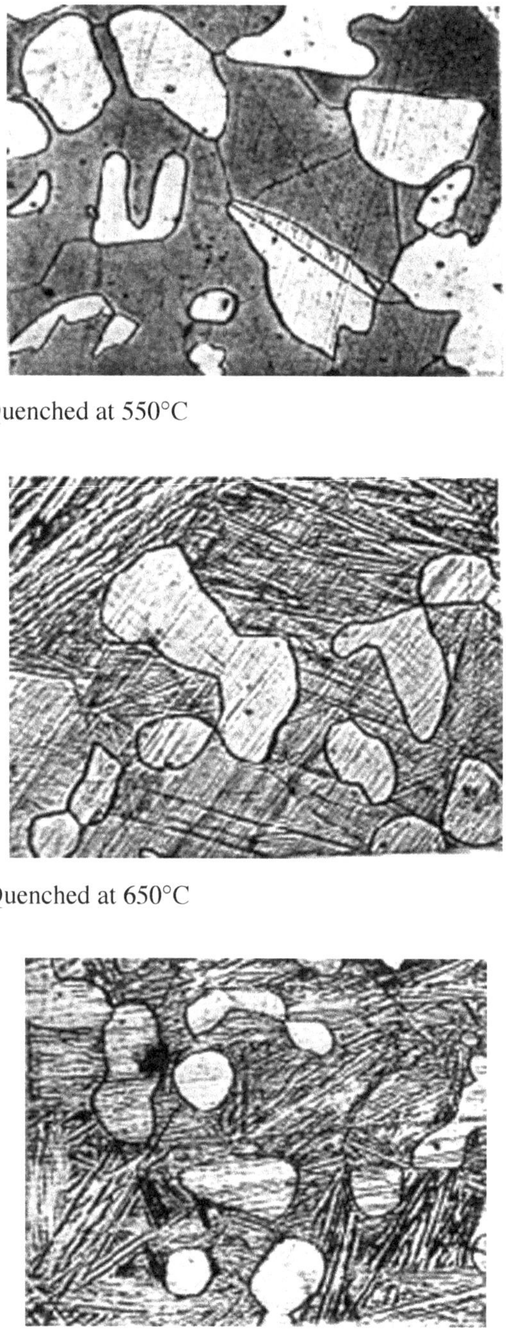

a: Quenched at 550°C

b: Quenched at 650°C

c: Quenched at 750°C

Figure 3.20 Photomicrographs of bronzes containing 22.27% Sn, quenched at different temperatures (a-c) (after Matsuda 1928, plates XV and XVI)

It is well known that lead is immiscible in copper (Scott 1991, 23) and is usually present as small spherical or dendritic droplets at the grain boundaries (Gettens 1969, 191). This agrees with casting experiments conducted by Staniaszek and Northover (1983, 263) and is also confirmed by the experimental casting of bronzes with a low lead content (2% Pb) reported here (fig. 3.21). The experiments have shown, furthermore, that in the bronzes with a high lead content (e.g. 10% Pb), lead was also present within the grains (fig. 3.22).

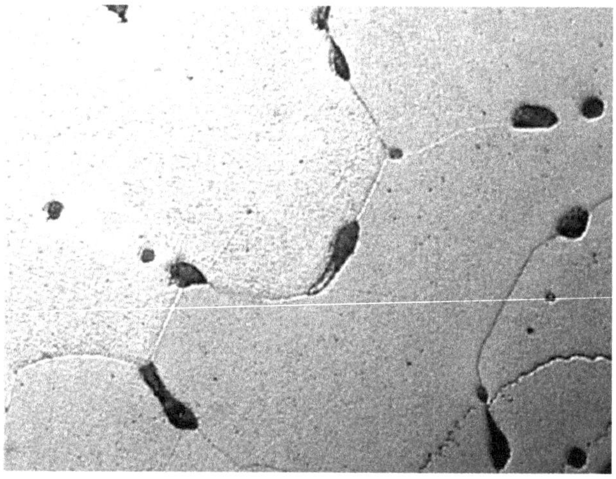

Figure 3.21 Photomicrograph of 2% Sn + 2% Pb bronze; water-quenched, showing lead droplets on the grain boundaries (image width 0.13 mm)

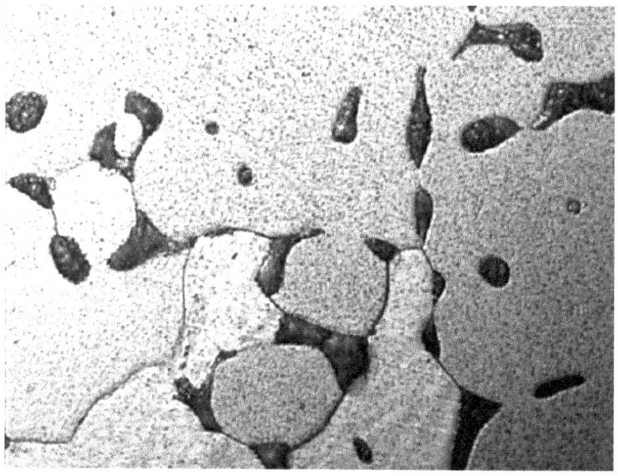

Figure 3.22 Photomicrograph of 2% Sn + 10% Pb bronze; water-quenched, showing lead droplets both on the grain boundaries and within the grains (image width 0.13 mm)

3.4.2 Grain size and dendritic arm spacing

Grain size and dendritic arm spacing were measured as described in chapter 2.3. The measurements are listed in table 3.2 and plotted in fig. 3.23. The figure shows that grain sizes vary dramatically according to composition and are not consistently correlated with cooling rate. On the other hand, dendritic arm spacing shows a clear correlation with cooling rate. Water-quenched cast bronzes consistently have smaller dendritic arm spacing than air-cooled ones.

The grain size cannot be smaller than the dendritic arm spacing. It is independent of solidification time. It is influenced by a large number of factors such as nucleation events on the side walls of the mould, foreign nuclei, e.g. inclusions, and chance events of damage or fragmentation from a variety of causes. Dendritic arm spacing, on the other hand, is controlled simply by cooling time (Campbell 1991, 146).

Table 3.2 ***Grain size and dendritic arm spacing of the bronzes cast in sand moulds***

Composition (wt%)			Dendritic arm spacing (μm)		Grain size (μm)	
Cu	**Sn**	**Pb**	**air-cooled**	**water-quenched**	**air-cooled**	**water-quenched**
98	2	0	—	—	175	378
94	6	0	60	55	1,020	860
90	10	0	46	38	460	120
85	15	0	40	33	925	1,580
77	23	0	—	—	894	555
96	2	2	50	35	625	176
92	2	6	57	40	140	207
88	2	10	—	—	450	209
92	6	2	40	33	392	900
88	6	6	33	30	200	330
84	6	10	40	30	330	250
88	10	2	30	29	250	500

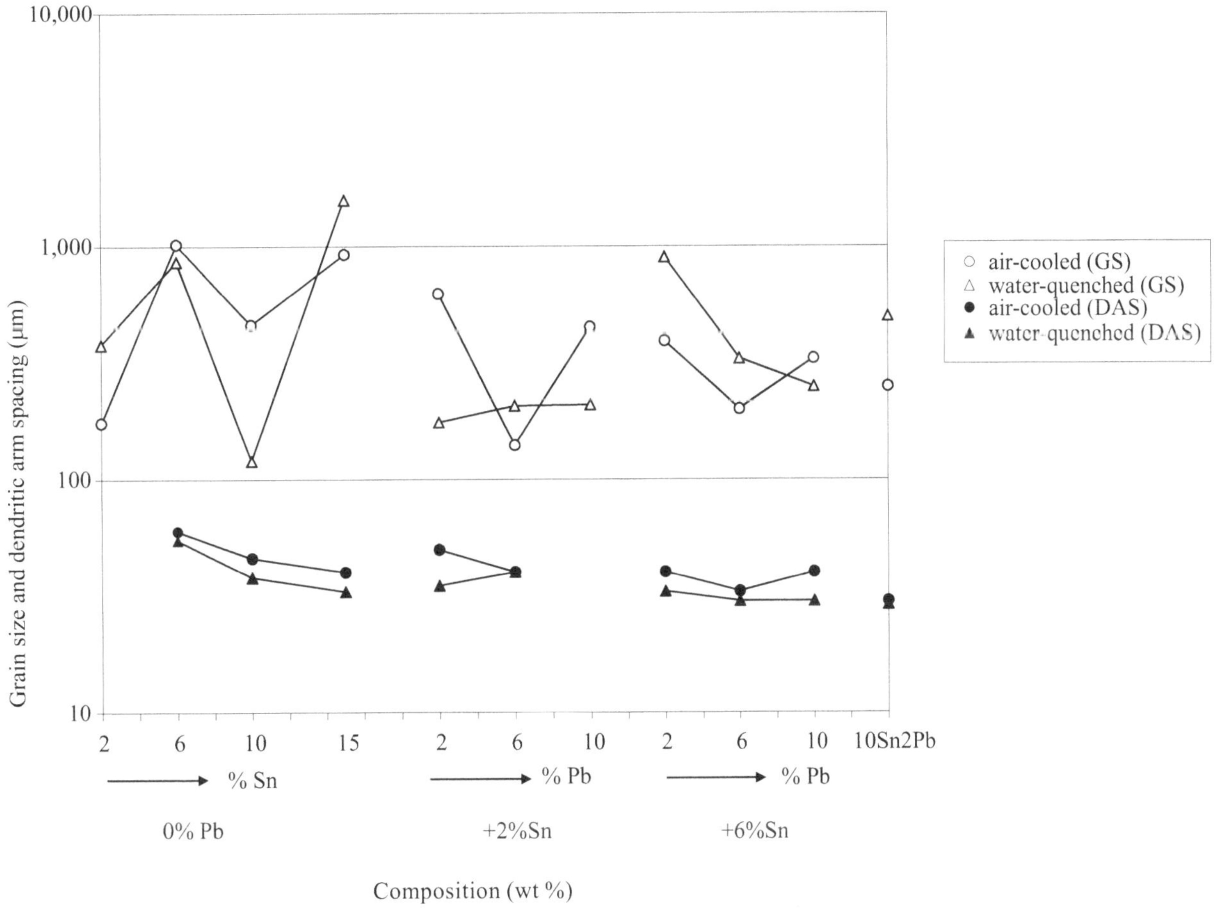

Figure 3.23 Grain sizes (GS) and dendritic arm spacing (DAS) of bronzes cast in sand moulds, showing a clear correlation between the dendritic arm spacing and the cooling rates.

3.4.3 Microhardness

The microhardness measurements, taken as discussed in chapter 2.3, are listed in table 3.3. The relationship of the microhardness of various compositions to different cooling rates is shown in figure 3.24 and reflects the well-established pattern of increased hardness with increasing amounts of tin. The addition of lead to a tin bronze, on the other hand, does tend to decrease hardness values slightly. Figure 3.24 also shows that most water-quenched cast bronzes have a slightly higher microhardness than air-cooled bronzes.

Table 3.3 ***Microhardness of bronzes cast in sand moulds***

Composition (wt %)			Air-cooled		Water-quenched	
Cu	**Sn**	**Pb**	**Hv**	**SD**	**Hv**	**SD**
98	2	0	62.9	1.0	64.1	2.0
94	6	0	77.7	6.8	74.8	7.2
90	10	0	99.9	4.4	104.0	5.7
85	15	0	126.8	21.5	153.0	24.1
77	23	0	249.4	20.3	263.6	25.7
96	2	2	58.5	2.0	57.9	1.9
92	2	6	54.3	10.2	57.1	6.5
88	2	10	54.5	3.6	55.2	2.2
92	6	2	74.5	8.4	84.2	5.5
88	6	6	69.6	1.7	68.0	3.7
84	6	10	68.4	5.7	78.6	6.9
88	10	2	91.8	6.0	105.1	11.1

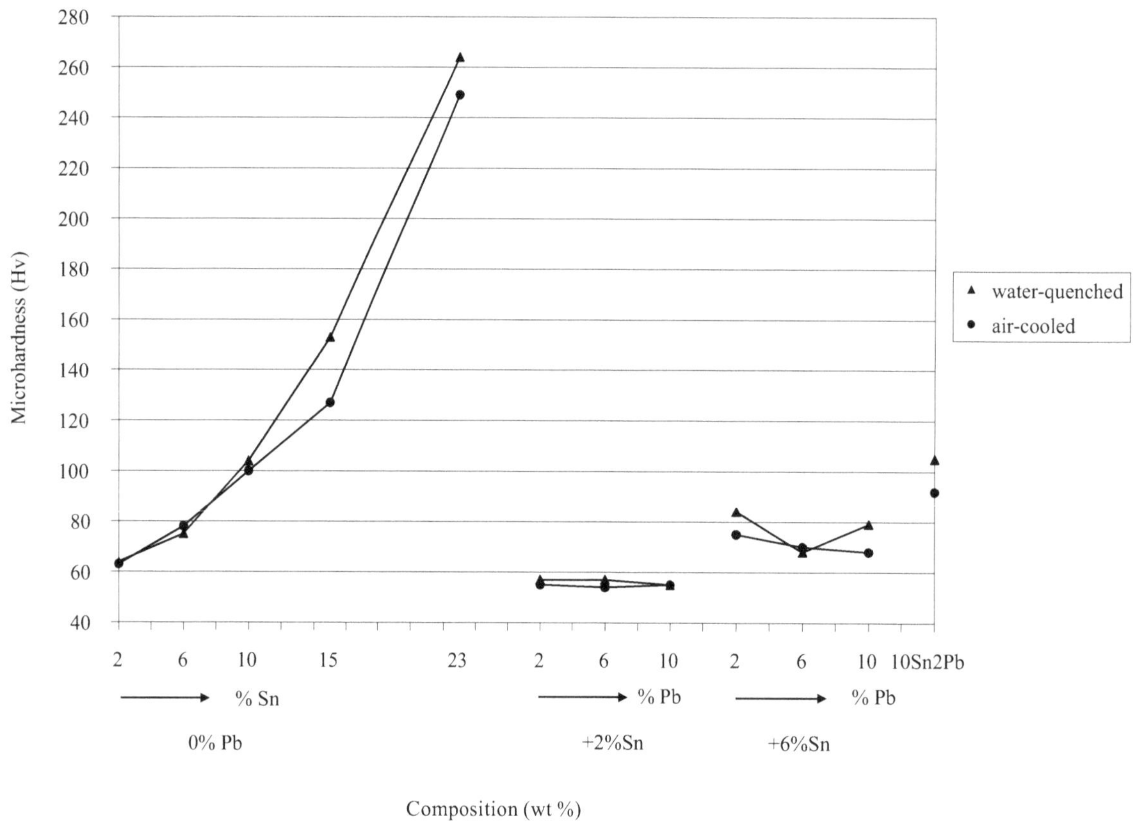

Figure 3.24 Microhardness of bronzes cast in sand moulds, showing that water-quenched bronze has a higher value than air-cooled bronze.

3.5 Conclusions

Twenty-four bronze castings, of 12 alloys and with two cooling rates for each alloy, have been carried out under controlled conditions. The results show that great efforts have to be made at each step, such as mould-making, melting the metal and pouring it into the moulds, to ensure a high-quality cast.

The low-tin (2% Sn) bronzes, including leaded ones, have only a granular structure or a mixed structure of dendrites and grains. Other cast bronzes with a higher tin content have dendritic structures. The microstructure of the water-quenched casts is appreciably different from that of the air-cooled bronzes with the same composition. The δ phase was seen in the water-quenched bronzes with 6% tin, while it was rarely seen in the air-cooled bronzes of the same alloy. The δ phase in the water-quenched casts looks abnormal in colour and appearance. The β phase began to appear in the water-quenched bronzes with a tin content of 15% and higher. This is much lower than that indicated by published equilibrium phase diagrams. The presence of δ and β phases in bronzes with such a low tin content seems to be peculiar to ancient artefacts.

A clear correlation between cooling rate and dendritic arm spacing was found, whereas grain size does not show such a clear correlation with cooling rate.

Microhardness values were dependent mainly on the composition of the metal, although almost all the water-quenched casts in the experimental series were found to be harder than the air-cooled casts.

4 Casting and metallography of bronzes cast in clay moulds

4.1 Introduction

Clay moulds for a whole range of tools and ornaments have been found in many Bronze Age contexts in Europe. It is possible that such moulds were in use even earlier, although no archaeological evidence has yet been uncovered. Observations have shown that clay moulds, tempered with chaff, fired at 700 and 900°C and used for casting bronzes, disintegrate into unrecognisable shapes after being left in the open for as little as one year. On the other hand, moulds with little organic temper, fired at 700°C and used for casting bronzes, remained reasonably intact and could still be recognised as moulds after being left exposed to the British weather for one year (Ottaway 2003, 345). The use of chaff might explain why so few clay moulds have been discovered in pre-Bronze Age contexts. Clay-moulding techniques were very sophisticated in ancient China and clay moulds were predominant throughout the Chinese Bronze Age. Bivalve moulds of clay were found, for instance, at Erlitou, a state-level settlement site that included palatial buildings, royal tombs and bronze foundries, and that dates to *c.* 1900 BC. By about 1500 BC piece-moulds were used in China. In Thailand bivalve ceramic (fired clay) moulds of large socketed axes and arrow points have been excavated at the Non Pa Wai production site dated to between 1500 and 700 BC (Pigott 1999, 16).

In Europe, the archaeological evidence seems to suggest that stone moulds were dominant during the Chalcolithic and the Early Bronze Age (Tylecote 1986, 84). Studies by Staniaszek and Northover (1983, 265) have shown that tin bronzes cast in stone or metal moulds were much easier to work than those cast in clay moulds. The question arises as to why stone moulds, which could be reused many times, were replaced by less durable clay moulds.

It has been reported that in many cases the clay used for Bronze Age bivalve moulds consisted of two grades: a well-prepared fine clay used for the inner part and a minimally prepared coarse one for the outer part (Collins 1970; Hodges 1958-9, 132; Tylecote 1986, 89). The fine inner layer allowed the accurate transfer of decoration from the mould to the cast artefact and the coarse outer layer wrapped the two halves together. Evidence of a continuous wrap around the joints securing the prepared inner valves in their correct position has been reported in Late Bronze Age mould fragments from Devon (Needham 1980, 181). However, two grades of clay for moulds were not always employed: moulds with a homogeneous clay body have been found in prehistoric Irish contexts (Coghlan and Raftery 1961, 243).

The majority of clay moulds found in European Bronze Age sites are bivalve moulds. The registration of the two halves was by means of keys (knobs and pits) (Coghlan 1960, 156; Tylecote 1986, 89). Indentations made using fingertips or small pebbles in the first piece automatically impressed small lugs in the second piece (Hodges 1958-9, 132). Nevertheless, keys for registration were not always used in Bronze Age or later moulds. For instance, no dowel-holes for location were seen in the Bronze Age bivalve moulds from Knighton, Devonshire (Hodges 1960, 153), and much later, during the Migration Period in Sweden, it was reported that dowels and holes for registration were extremely rare in moulds used for the manufacture of jewellery (Lamm 1973, 3).

Analytical data on clay moulds are rare in the literature. However, it has generally been suggested that the moulds were made of clay with the addition of sand and vegetable matter such as straw or dung (Tylecote 1986, 84; Hodges 1958-9, 132). Clay moulds used in Chinese bronze castings were also of a sand-clay-fibre mixture, which permitted gas to escape during casting (Gettens, 1969, 31). If a core was used for producing a hollow artefact, it usually had more quartz and coarser grains than the mould (Hua Jueming 1985, 483). Recent studies of Chinese clay mould materials and experiments carried out by Tan Derui and Huang Long (1996, 47-70) of the Shanghai Museum revealed that local earth mixed with sand and plant ash was used for mould production. The chemical compositions of a few clay moulds are listed in table 4.1. The addition of sand, which contains mainly quartz, can reduce contraction and improve refractability. The addition of organic materials, e.g. straw, dung, or plant ash, can increase the permeability, reduce the thermal accumulation coefficient of the clay mould and improve the ability of the molten metal to fill the mould (Tan Derui and Huang Long 1996, 67).

Table 4.1 *Chemical compositions of archaeological clay moulds*

Sample	Origin and date	Na_2O	MgO	Al_2O_3	SiO_2	K_2O	CaO	TiO_2	FeO	Reference
Mould of vessel	Houma, China 500 BC	1.6	1.7	11.0	63.1	3.8	13.3	0.6	4.0	Wang Quanyu 2002, 32
Mould of tool, brown	Houma, China 500 BC	1.9	1.7	14.0	63.3	3.7	10.4	0.6	4.4	Wang Quanyu 2002, 32
Mould of tool, grey	Houma, China 500 BC	2.2	1.7	12.8	63.6	3.9	10.8	0.5	4.5	Wang Quanyu 2002, 32
Mould of vessel, black	Fufeng, China 900 BC	2.4	1.5	11.4	68.4	4.3	7.0	0.7	4.4	Wang Quanyu 2002, 32
Mould of vessel, grey	Fufeng, China 900 BC	1.7	1.8	10.8	72.5	3.1	5.6	0.4	4.1	Wang Quanyu 2002, 32
Mould O Mould K	Athens, Greece 500 BC	1.4 0.2	2.4 1.6	10.8 11.9	62.3 61.6	1.7 2.1	15.9 16.8	0.6 0.8	4.6 4.7	Schneider 1989, 306

Clay moulds were fired at a temperature that reduced the gas content in the clay and resulted in greater strength and better thermal properties, thus producing high-quality bronzes. It has been reported that most European clay moulds were fired to a temperature of about 650°C (Tylecote 1986, 89; Needham 1980, 192). Estimates of the firing temperature were obtained from refiring experiments. However, the accuracy of the estimates is doubtful, as shrinkage can be too small to be measured for clay fired at a temperature lower than its softening point during refiring. This has been proved by Andrews (pers. comm.) in his refiring experiments with controlled samples. Providing the estimated firing temperature of 650°C is correct, the clay for the mould must have been a non-calcareous clay, as calcite starts to decompose at 650°C and the decomposition is not complete until the temperature reaches 900°C. Moulds made of calcareous clay had to be fired to a temperature at which the decomposition of calcite had almost finished to ensure the complete escape of carbon dioxide from the mould and produce a high-quality casting. Chinese moulds from Houma were estimated to have been fired at 850-920°C (Tan Derui and Huang Long 1996, 55), a temperature at which calcite has decomposed almost completely and stopped producing carbon dioxide.

The thickness of the clay mould is also important. If the mould is too thin, it is not strong enough to withstand having molten metal poured into it; on the other hand, if it is too thick, it takes a long time to dry and easily breaks during firing if it is not completely dry. Most of the European Bronze Age clay moulds were about 1-2 cm thick (Coghlan and Raftery 1961, 240; Collins 1970; Needham 1980, 197-9).

4.2 Mould-making

Industrial fireclay with amounts of SiO_2 similar to those of the moulds listed in table 4.1 was chosen to make the clay moulds for the experimental series reported here. Its chemical composition is shown in table 4.2

Table 4.2 ***Chemical composition of clay used for moulding (wt %)***

SiO_2	TiO_2	Al_2O_3	Fe_2O_3	CaO	MgO	K_2O	Na_2O	Others
65.00	0.60	24.10	1.70	1.10	0.5	0.80	0.86	5.34

Because of the high content of SiO_2 in the clay, no sand was added. Sawdust was added to increase the permeability of the moulds. Sawdust was thought to be the easiest and most pleasant material to work with compared with straw, dung, or plant ash. Three different mixtures of clay and sawdust, containing 10%, 20% and 30% sawdust by weight, were used in a trial. The results showed that clay with 30% sawdust has too little plasticity to be used for moulding, while clay with 10% sawdust was not permeable enough to make a good-quality casting. Clay with 20% sawdust was therefore chosen to make all the clay moulds for this series of experiments. The sawdust was sieved through a 1 mm sieve before it was added to the clay.

In the trials preceding the experimental series, the first half of the mould was made and air-dried; the pattern was left in place and the second half made to fit the first one when the first half was leather-hard. It was found that a precise registration of the second piece was impossible with this method, because the first half of the mould had shrunk during drying and the second half shrank and became smaller than the first piece. The linear shrinkage of the moulds was approximately 5%.

After several trials it was found that a good registration could be achieved by making the two halves of the mould at the same time. This was carried out as follows: two wooden frames (fig. 4.1) were used to hold the clay, the plasticity of which was reduced by the addition of sawdust, making it difficult to shape. For the bottom half of the mould, one frame was filled with clay and the wooden pattern was pressed into the clay. Sawdust was sieved over this half of the mould to ease the separation of the two halves of the mould at a later stage (fig. 4.2). The top half of the mould was prepared by pressing clay into the second frame, which had been placed on top of the first frame. The two halves were then separated (fig. 4.3). A runner inlet was made next to the cavity in the mould and three small holes were drilled through the mould from the cavity to aid the escape of gases. The frames and the pattern were removed (fig. 4.4). The two halves were then placed carefully one on top of the other and the whole assemblage was sealed around the sides with the same mixture of clay and sawdust as that used for the mould. A clay runner bush was added to complete the assemblage (fig. 4.5). The moulds were air-dried overnight and then put into a drying oven to dry at 40-50°C for two days.

To test the effect of the firing temperature on the properties of the moulds and on the bronzes cast within them, two moulds were fired at 700 and 900°C respectively. Little difference was observed in either moulds or castings, which is probably due to the non-calcareous nature of the clay (see section 4.1). Since the sawdust had burnt out at about 500°C, leaving pores for gases to escape during casting, it was decided to use the lower firing temperature of 700°C for the series of experiments with clay moulds. This temperature was reached by the following regime: heating to 100°C at a rate of 100°C/hour, resting for two hours to allow the moisture to evaporate completely and then heating to 700°C at the same rate of 100°C/hour. After resting for one hour the moulds were allowed to cool down inside the furnace. The moulds were then ready for the casting experiments (fig. 4.6).

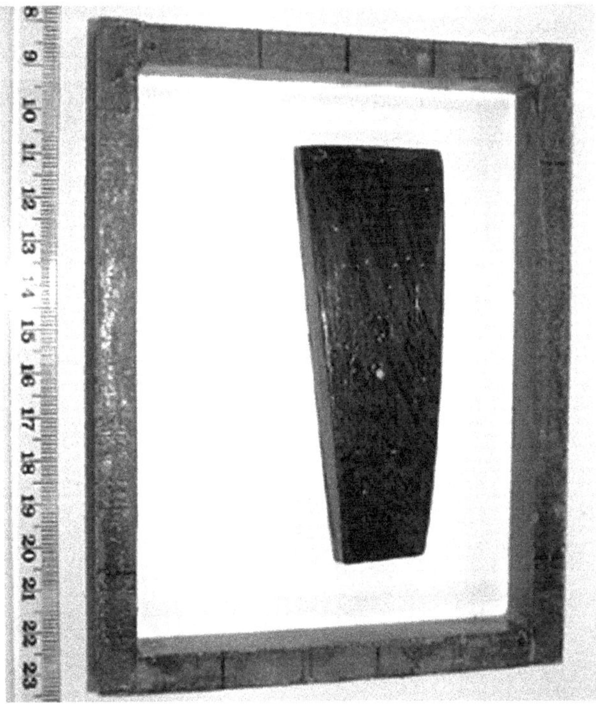

Figure 4.1 The pattern and frame used to prepare the mould

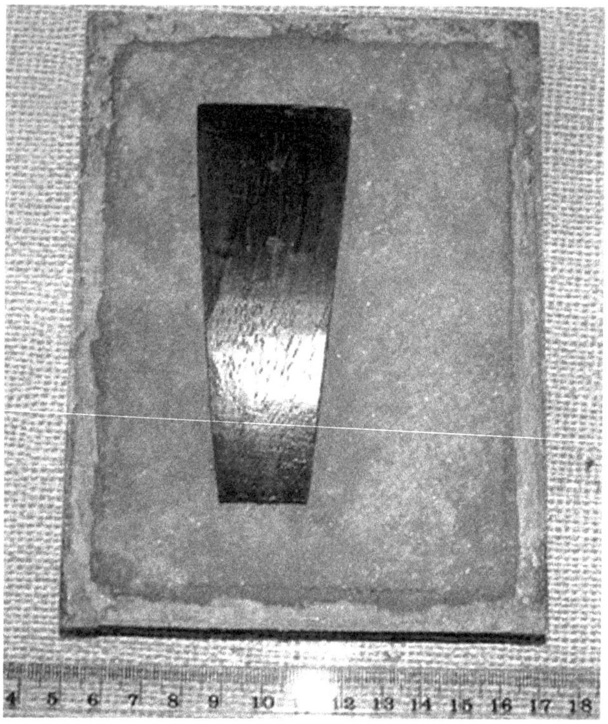

Figure 4.2 The completed first half of the mould dusted with sawdust

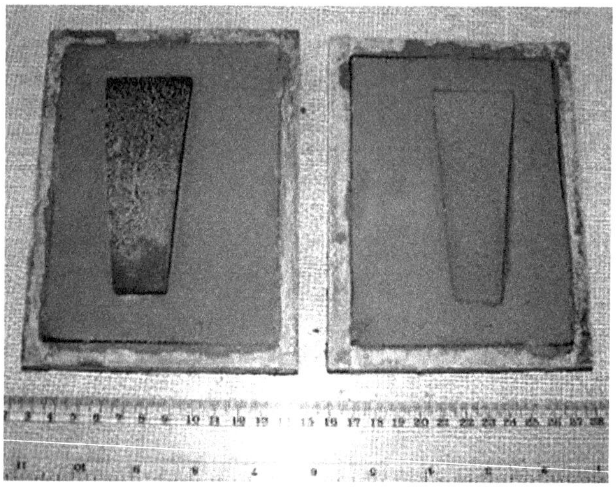

Figure 4.3 The two halves of the mould separated

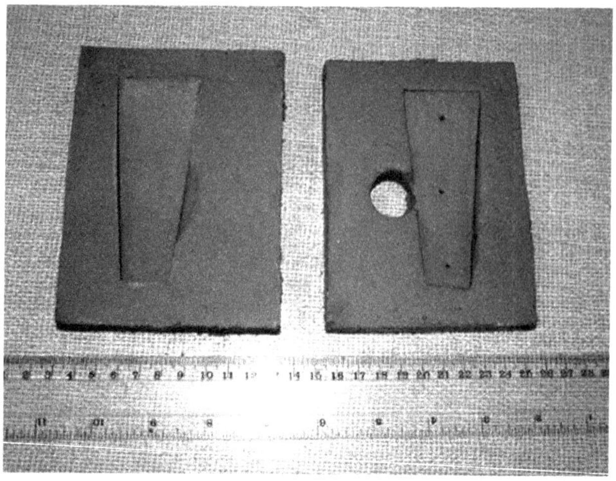

Figure 4.4 Two completed parts of the mould, showing small holes and runner inlet

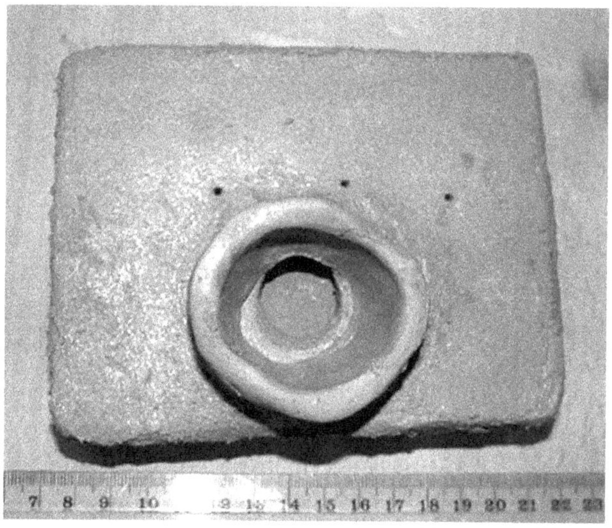

Figure 4.5. A completed mould assemblage with runner bush in place

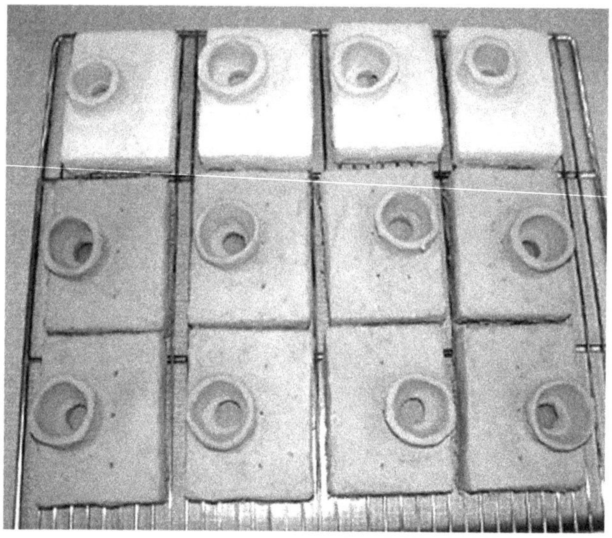

Figure 4.6 Dried moulds (two bottom rows) and fired moulds (top row)

4.3 Experimental casting

In initial experiments, the moulds were tied together with leather strips to prevent the molten metal from running out. However, as the leather strips were burnt and broken when the molten metal was being poured, the moulds were not tied together in subsequent experiments. Extensive flashing occurred in all initial castings (fig. 4.7). This could be partly due to further shrinkage of the clay and partly to the pressure of the molten metal, pushing the two pieces of the mould apart. To reduce flashing, heavy pieces of metal were put on top of the mould to weight it down (fig. 4.8).

A series of 36 castings, on material with a total of 12 different compositions and three different cooling rates for each composition, was carried out. The moulds were prepared for the casting in three different ways: one was preheated (to 350°C) and air-cooled; one was not preheated and was air-cooled; the third was not preheated but was water-quenched. This was done by breaking the mould and immersing the object in water within a couple of minutes of pouring. The casting conditions for each cast are listed in table 4.3. All cast artefacts were sectioned for metallographic observation and microhardness testing. The results are presented in the next section.

Table 4.3 ***Composition, cooling regime and microstructure of the bronze axes cast in the clay moulds***

Sample no.	Composition[1] (wt %)		Cooling method	Microstructure	Figure no.
	Sn	Pb			
111	2	—	Preheated, air-cooled	Granular, no α+δ eutectoids	—
112	2	—	Unpreheated, air-cooled	Granular, no α+δ eutectoids	4.14
113	2	—	Unpreheated, water-quenched	Granular, α+δ eutectoids on grain boundaries	4.15
114	6	—	Preheated, air-cooled	Dendritic, very few α+δ eutectoids	4.11
115	6	—	Unpreheated, air-cooled	Dendritic, very few α+δ eutectoids	4.12
116	6	—	Unpreheated, water-quenched	Dendritic, α+δ eutectoids in the interdendritic regions, no β	4.13
117	10	—	Preheated, air-cooled	Dendritic, α+δ eutectoids can be clearly seen	—
118	10	—	Unpreheated, air cooled	Dendritic, α+δ eutectoids can be clearly seen	—
119	10	—	Unpreheated, water-quenched	Dendritic, β infilled between dendrites	4.16 and 4.17
120	15	—	Preheated, air-cooled	Dendritic, very many α+δ eutectoids	—
121	15	—	Unpreheated, air-cooled	Dendritic, very many α+δ eutectoids	4.18
122	15	—	Unpreheated, water-quenched	Dendritic, β infill between dendrites and within α grains	4.19
123	23	—	Preheated, air-cooled	Dendritic, very many α+δ eutectoids	—
124	23	—	Unpreheated, air-cooled	Dendritic, very many α+δ eutectoids	4.20
125	23	—	Unpreheated, water-quenched	β matrix with islands of α and δ	4.21-4.23
126	2	2	Preheated, air-cooled	Dendritic, no α+δ eutectoids	—
127	2	2	Unpreheated, air-cooled	Dendritic, no α+δ eutectoids	—
128	2	2	Unpreheated, water-quenched	Dendritic, dark phase in the interdendritic regions (α or δ? See table 4.4 sample no. 128)	4.27 and 4.28
129	2	6	Preheated, air-cooled	Granular, no α+δ eutectoids, Pb on grain boundaries	4.24
130	2	6	Unpreheated, air-cooled	Dendritic, no α+δ eutectoids	4.25
131	2	6	Unpreheated, water-quenched	Dendritic, no α+δ eutectoids	—
132	2	10	Preheated, air-cooled	Granular, no α+δ eutectoids, Pb both on grain boundaries and within the grains	—

Table 4.3 (contd.) ***Composition, cooling regime and microstructure of the bronze axes cast in the clay moulds***

133	2	10	Unpreheated, air-cooled	Granular, no α+δ eutectoids, Pb both on grain boundaries and within the grains	4.26
134	2	10	Unpreheated, water-quenched	Granular, no α+δ eutectoids, Pb both on grain boundaries and within the grains	—
135	6	2	Preheated, air-cooled	Dendritic, very few α+δ eutectoids	—
136	6	2	Unpreheated, air-cooled	Dendritic, very few α+δ eutectoids	—
137	6	2	Unpreheated, water-quenched	Dendritic, eutectoids with dark δ phase in the interdendritic regions	
138	6	6	Preheated, air-cooled	Dendritic, very few α+δ eutectoids	—
139	6	6	Unpreheated, air-cooled	Dendritic, very few α+δ eutectoids	—
140	6	6	Unpreheated, water-quenched	Dendritic, eutectoids with dark δ phase in the interdendritic regions	—
141	6	10	Preheated, air-cooled	Dendritic, very few α+δ eutectoids	—
142	6	10	Unpreheated, air-cooled	Dendritic, very few α+δ eutectoids	—
143	6	10	Unpreheated, water-quenched	Dendritic, eutectoids with dark δ phase in the interdendritic regions	
144	10	2	Preheated, air-cooled	Dendritic, α+δ eutectoids in the interdendritic regions	—
145	10	2	Unpreheated, air-cooled	Dendritic, α+δ eutectoids in the interdendritic regions	—
146	10	2	Unpreheated, water-quenched	Dendritic, eutectoids with dark δ phase in the interdendritic regions; β possibly present within the eutectoid	—

Note: [1] The balance is copper.

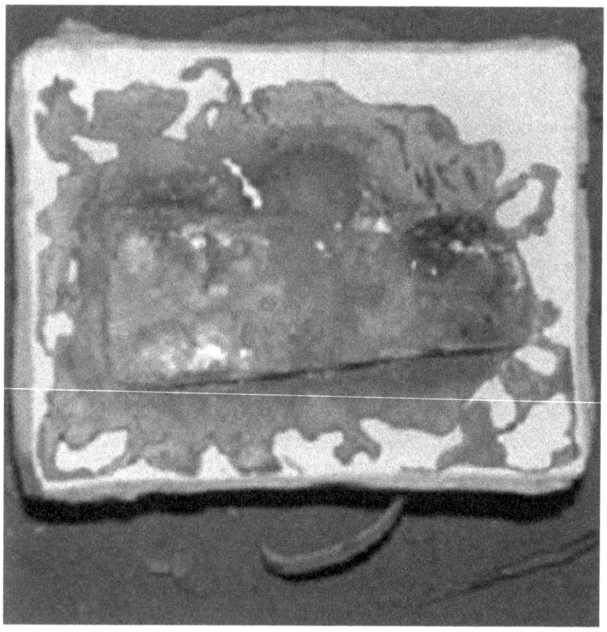

Figure 4.7 Extensive flashing in an initial casting experiment

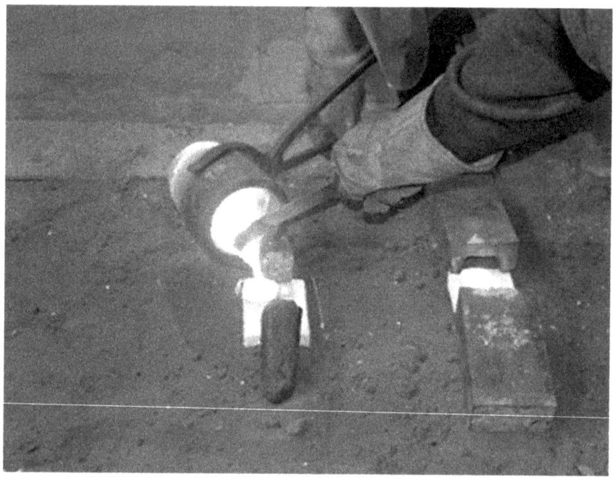

Figure 4.8 Pouring molten metal into a clay mould, weighted down by heavy metal pieces to prevent excessive flashing

4.4 Results and discussion

4.4.1 Macrostructure

Axes cast in clay moulds have a finer surface than those cast in sand moulds (fig. 4.9). This can be attributed to the small grain size of clay, which is generally in the range of 5-50 µm. Shrinkage was sometimes considerable (fig. 4.10). This is probably due to rapid cooling of the surface where molten metal has solidified too quickly to allow the space caused by shrinkage to be filled from the reservoir in the runner bush.

Flashing was almost unavoidable in clay-mould castings. The actual size of the cast axes was smaller than the pattern because of the shrinkage of the clay.

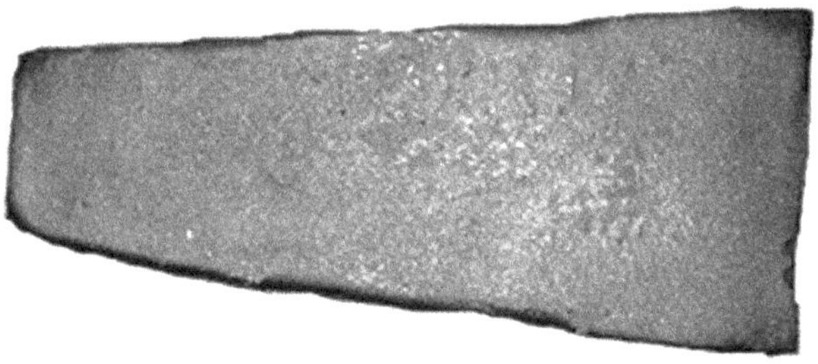

Figure 4.9 Sand-blasted axe (6% Sn+10% Pb) cast in clay mould, showing smooth surface

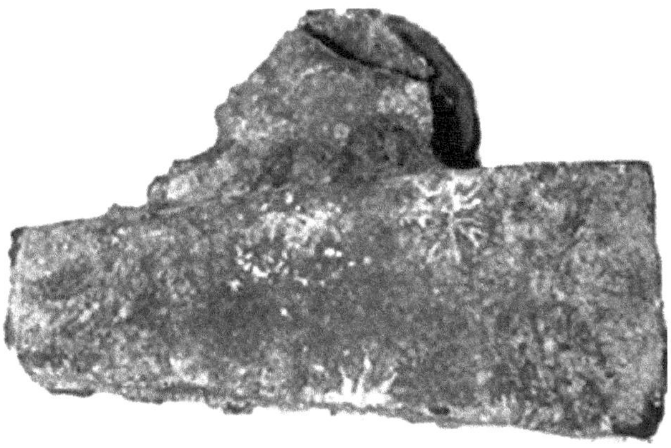

Figure 4.10 Untreated axe (23% Sn) cast in clay mould, showing shrinkage on the surface

4.4.2 Microstructure

Metallographic examination has shown that some bronzes cast in clay moulds have a granular structure while others have a dendritic structure, but some casts have a mixed structure of grains and dendrites. Figures 4.11–4.28 present a selection of photomicrographs in black and white in this volume. They are shown again in colour in appendix 1 on the accompanying CD. A complete series of photomicrographs of bronzes cast in clay moulds is given in appendix 3 on the CD.

In general, little difference in the microstructures of bronzes cast in preheated moulds and those cast in unpreheated moulds could be found. In the 6% tin bronzes, for instance, both have a dendritic structure and have a few islands of eutectoids ($\alpha+\delta$) (figs. 4.11 and 4.12). On the other hand, water-quenched bronzes are very different from air-cooled ones. The water-quenched 6% tin bronze, for instance, has, in contrast to the air-cooled bronze, a very pronounced dendritic structure, smaller grain size and many interdendritic eutectoids ($\alpha+\delta$) on the grain boundaries in which the δ phase is darker than in the air-cooled bronze (fig. 4.13). Bronzes with 2% tin, cast in preheated and unpreheated moulds and air-cooled, do not have a δ phase at all and generally have a granular structure (fig. 4.14). However, the water-quenched cast of the same composition has some $\alpha+\delta$ eutectoids on the grain boundaries (fig 4.15). The δ phase in the water-quenched cast looks dark under the microscope rather than light blue. Its appearance is also different from that of the air-cooled sample. The water-quenched 10% tin bronze has a pronounced dendritic structure with very many $\alpha+\delta$ eutectoids (fig. 4.16). At higher magnification an interdendritic needle β phase can be recognised (fig. 4.17). The air-cooled 15% tin bronze has very many interdendritic eutectoids ($\alpha+\delta$) (fig. 4.18), while the water-quenched 15% tin bronze has β phase both interdendritically and within the α grains (fig 4.19).

Bronze containing 23% tin is quite different from the other bronzes. The predominant phase of the air-cooled cast bronze with 23% tin is the δ phase (fig. 4.20). The matrix of the water-quenched bronze of the same composition, confirmed by SEM analysis (sample no. 125 in table 4.4), is needle β phase with islands of α phase and δ phase (figs. 4.21 and 4.22). The presence of a β rather than γ phase indicates that the quenching temperature was between 586 and 788°C as the γ phase is found in high-tin bronzes quenched at temperatures of 520-585°C (Chadwick 1939, 344; Goodway and Conklin 1987, 10). The γ phase, which tends to look a uniform greyish or light brownish colour on etching (Scott, pers. comm.), does not seem to be present in the cast bronzes discussed here. There is a dark phase at the grain boundaries (fig. 4.23). This could be either a γ or a δ phase. Its morphology suggests that it is more likely to be a δ phase (Scott, pers. comm.).

The microstructures of the leaded bronzes are even more complex (table 4.3). For example, the casts with 2% tin but varying lead contents, e.g. 2, 6 and 10% lead, have different microstructures. The cast bronzes with 2% tin and 2% lead all have a dendritic structure regardless of the cooling method. The bronze with 2% tin + 6% lead cast in a preheated mould and air-cooled, on the other hand, has a granular structure (fig. 4.24), while the ones cast in moulds at room temperature and both air-cooled and water-quenched have dendritic structures (fig. 4.25).

In bronzes cast in clay moulds, just like the bronzes cast in sand moulds, lead is usually present as small spherical or dendritric droplets at the grain boundaries (fig. 4.26). The lead globules are sometimes difficult to recognise in an etched sample (fig. 4.27). However, they can be seen clearly in a backscattered electron image (fig. 4.28).

In principle, lead should not affect the microstructure of the bronzes because it is immiscible in copper. However, it can improve the fluidity of the alloy in the melt because of its low melting point. This may explain why the microstructures of these leaded bronzes are different from those in the corresponding tin bronzes. The casting temperature could be the major factor. On the one hand, the addition of lead reduced the melting point of the metal; on the other hand, it lowered the casting temperature, because it was added to the molten copper just before it was poured into the mould. The combination of these effects may have altered the actual casting temperature, although an attempt was made to keep all the melting temperatures constant (see chapter 2).

Compositional analysis was carried out on two water-quenched samples with a JEOL JSM-6400 Scanning Electron Microscope equipped with EDS for confirmation of the metallographic observation. The operating condition was 20KV. The results are listed in table 4.4. These data show the distribution of alloying elements in different phases; they do not refer to the average composition of the cast object. However, the composition of the matrix (area analysis by SEM/EDS) can provide an estimate of the alloy composition. Table 4.4 shows that the compositional results from the ICP-AES analysis are quite close to the results produced by SEM/EDS. The SEM/EDS analyses also confirmed that the dark phase (in fig. 4.23), which is darker than the normal δ phase, is δ phase.

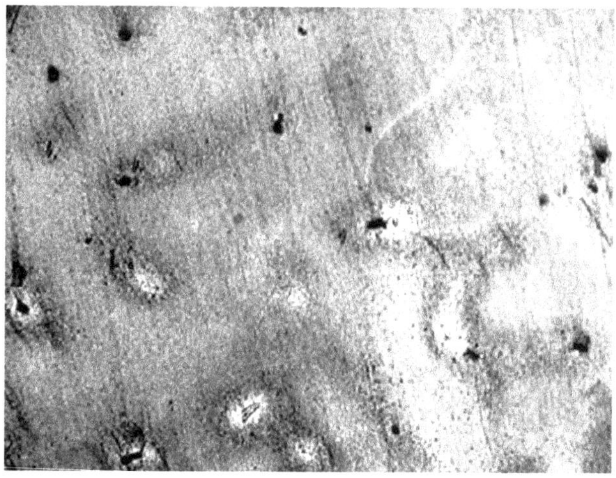

Figure 4.11 6% Sn bronze cast in a preheated (350°C) mould and air-cooled, showing very few islands of eutectoids (α+δ) (image width 0.33 mm)

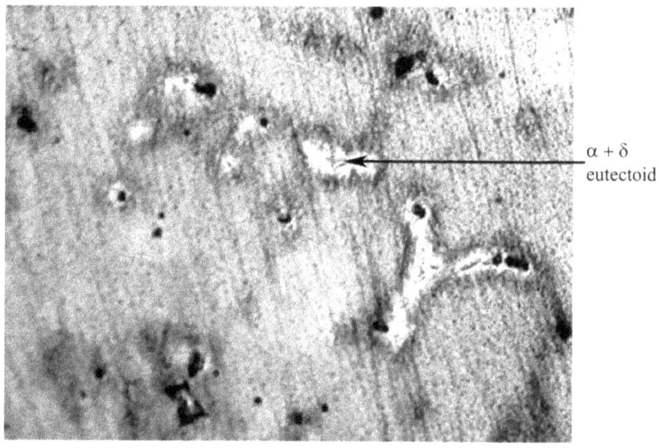

Figure 4.12 6% Sn bronze cast in an unpreheated mould and air-cooled, showing islands of eutectoids (α+δ) (image width 0.33 mm)

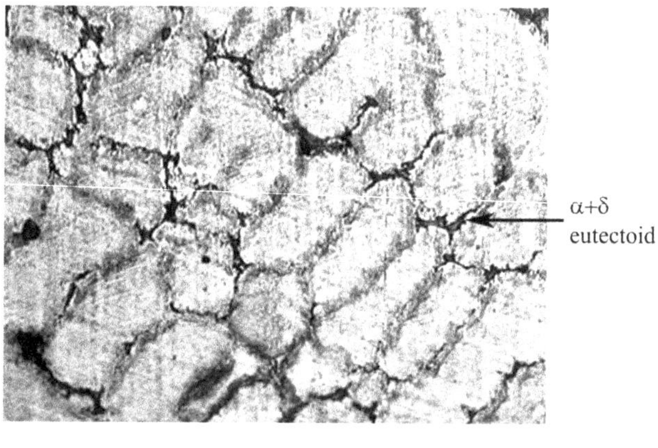

Figure 4.13 6% Sn bronze cast in an unpreheated mould and water-quenched, showing α+δ eutectoids on grain boundaries, where the δ phase is darker than usual (image width 0.33 mm)

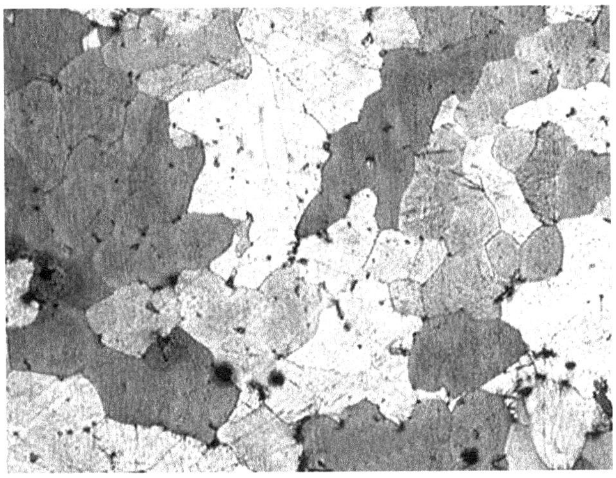

Figure 4.14 2% Sn bronze cast in an unpreheated mould and air-cooled, showing a granular structure (image width 1.3 mm)

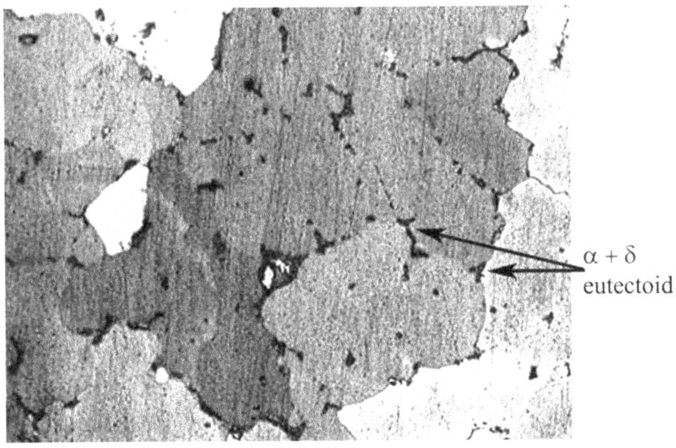

Figure 4.15 2% Sn bronze cast in an unpreheated mould and water-quenched, showing α+δ eutectoids on grain boundaries (image width 0.65 mm)

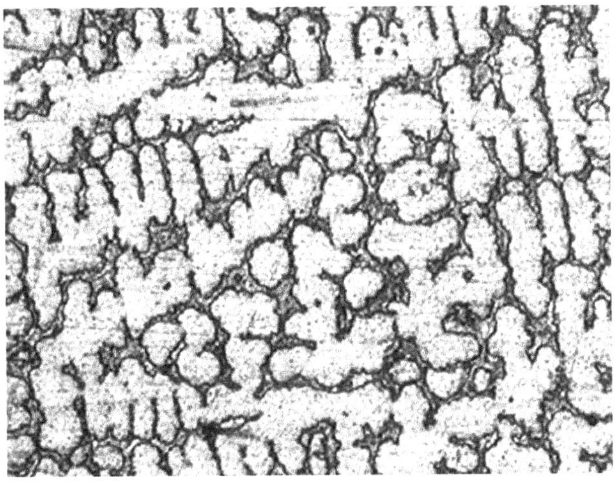

Figure 4.16 10% Sn bronze cast in an unpreheated mould and water-quenched, showing pronounced dendritic structure and very many eutectoids (α+δ) (image width 0.65 mm)

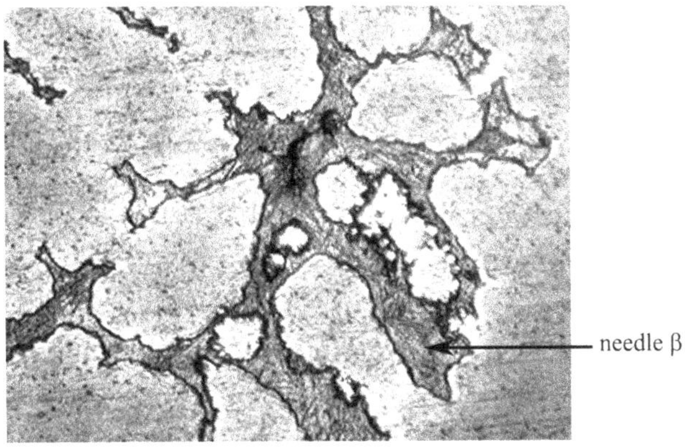

Figure 4.17 Same sample as figure 4.16 at higher magnification, showing interdentritic needle β phase (image width 0.16 mm)

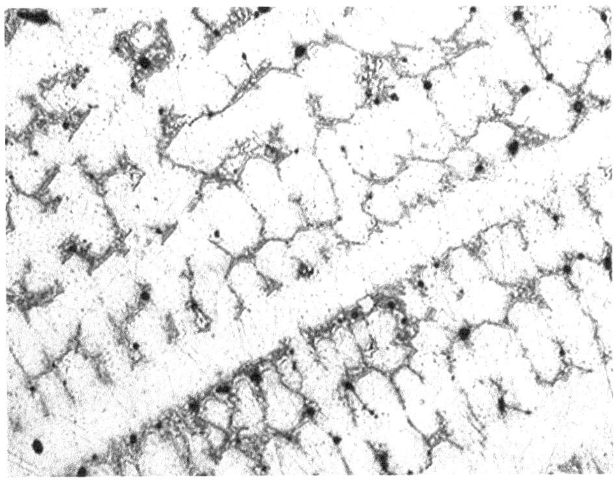

Figure 4.18 15% Sn bronze cast in an unpreheated mould and air-cooled, showing very many interdendritic eutectoid α+δ (image width 0.65 mm)

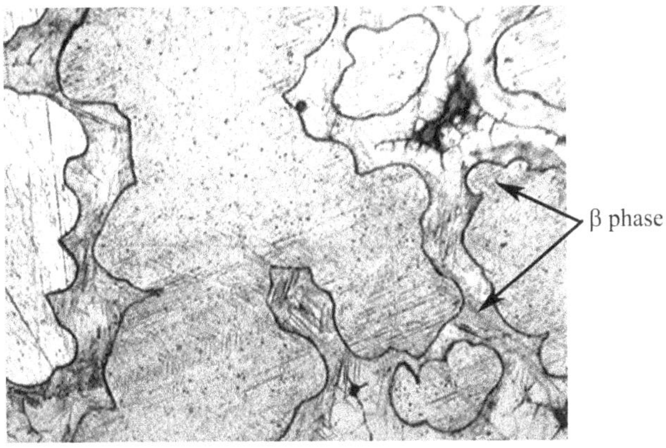

Figure 4.19 15% Sn bronze cast in an unpreheated mould and water-quenched, showing β phase both interdendritically and within α grains (image width 0.16 mm)

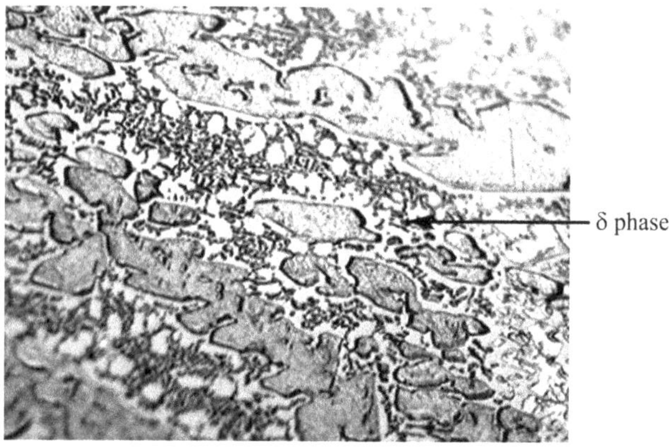

Figure 4.20 23% Sn bronze cast in an unpreheated mould and air-cooled, showing massive δ phase (image width 0.16 mm)

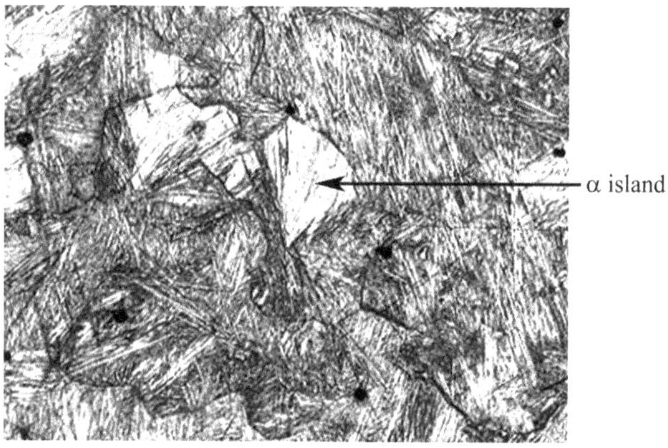

Figure 4.21 23% Sn bronze cast in an unpreheated mould and water-quenched (sample no. 125), showing α islands in the matrix of needle β phase (image width 1.3 mm)

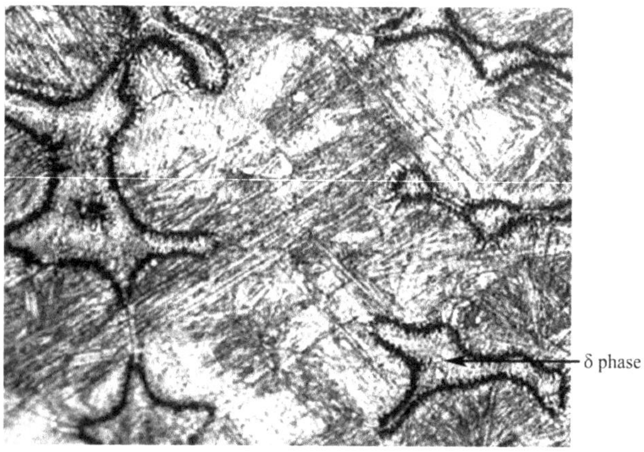

Figure 4.22 Same sample as figure 4.21, showing δ phase in a matrix of β phase (image width 0.33 mm)

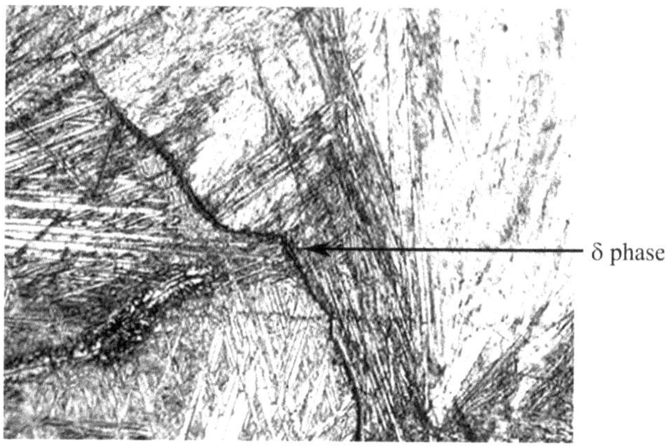

Figure 4.23 Same sample as figure 4.21, showing δ phase on the grain boundaries (dark line). This was confirmed by SEM/EDS analysis (image width 0.33 mm)

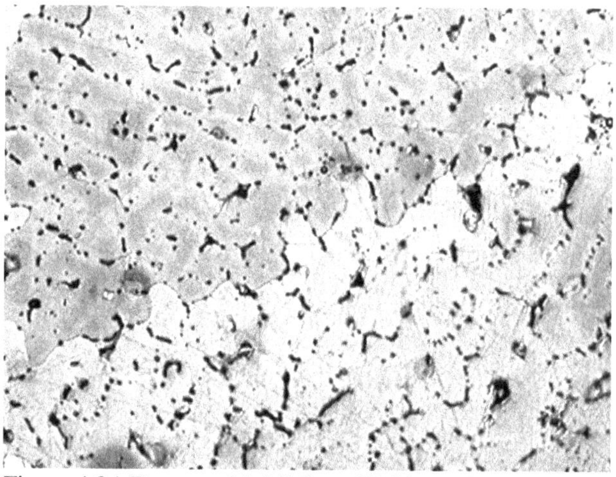

Figure 4.24 Bronze with 2% Sn + 6% Pb cast in a preheated mould and air-cooled, showing a granular structure (image width 1.3 mm)

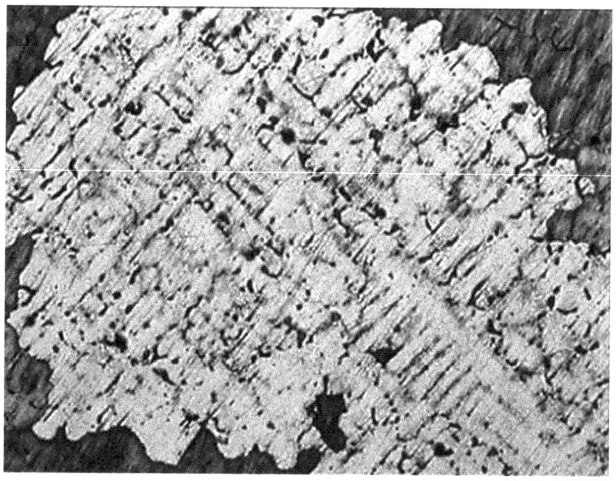

Figure 4.25 Bronze with 2% Sn + 6% Pb cast in an unpreheated mould and air-cooled, showing a dendritic structure (image width 1.3 mm)

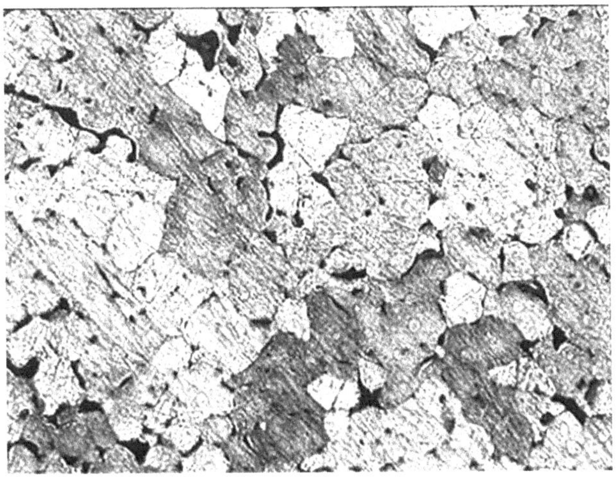

Figure 4.26 Bronze with 2% Sn + 10% Pb cast in an unpreheated mould and air-cooled, showing granular structure and Pb on grain boundaries (image width 0.65 mm)

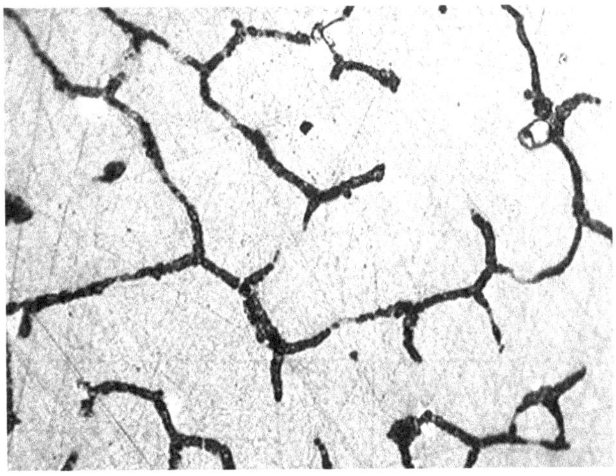

Figure 4.27 Bronze with 2% Sn + 2% Pb cast in an unpreheated mould and water-quenched (sample no. 128), showing a dendritic structure; Pb is almost indistinguishable from the dark phase in the interdendritic regions (image width 0.33 mm)

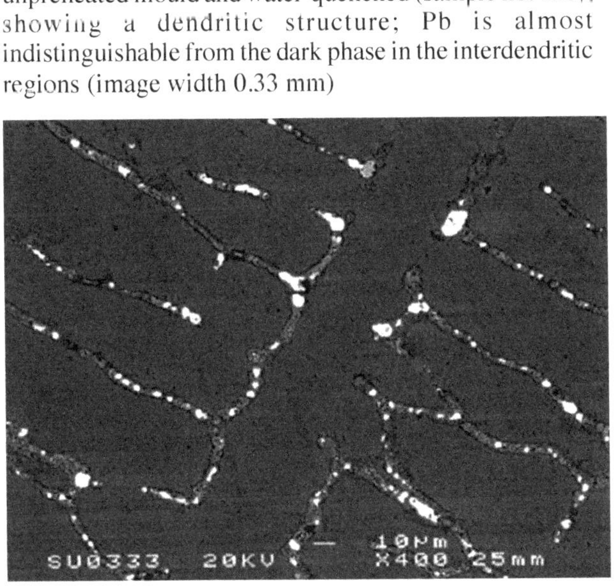

Figure 4.28 Backscattered electron image of the same sample as in figure 4.27, showing lead globules in the interdendritic phase

4.4.3 Grain size and dendritic arm spacing

Grain size and dendritic arm spacing were measured as described in chapter 2.3. The results of the measurement are listed in table 4.5 and plotted in figure 4.29. Grain sizes vary dramatically among these casts. There is a general trend towards larger grain size in bronzes cast in preheated moulds and smaller grain size in bronzes cast in clay moulds and then water-quenched, but there are notable exceptions, which do not show any correlation with cooling rate (see fig. 4.29 and table 4.5). On the other hand, dendritic arm spacing shows a clear correlation with cooling rate: water-quenched casts have a smaller dendritic arm spacing than casts cooled by the other two methods (fig. 4.29 and table 4.5).

4.4.4 Microhardness

The microhardness measurements, taken as discussed in chapter 2.3, are listed in table 4.6. The correlation of the microhardness of various compositions with different cooling rates is shown in figure 4.30. There is not much difference in microhardness between casts made in preheated moulds and those made in unpreheated moulds, but most of the water-quenched casts have higher microhardness values. The graphs also show that the addition of a small amount of lead, as such in the 6% Sn + 2% Pb and 10% Sn + 2% Pb bronzes, can increase the hardness of water-quenched casts considerably, although it is generally accepted that lead decreases the hardness of bronzes. It is interesting to note that, in their experiments with tin bronzes containing varying amounts of lead, Staniaszek and Northover (1983, 264) also found that bronzes with 2-4% lead had high microhardness values. It would be interesting to investigate this effect further.

The microhardness values for water-quenched bronzes containing 23% tin are, in contrast to the other bronzes investigated here, very little different from air-cooled ones. Yet most ancient high-tin bronzes were quenched (Goodway and Conklin 1987, 80; Meeks 1993, 75; Srinivasan 1998, 80). This can most probably be explained by the advantage given by the structure of the β phase. A bronze with a martensitic β needle phase, such as found in the water-quenched 23% tin bronze (figs. 4.21-4.23) is much less brittle than the same bronze cooled slowly to room temperature (Scott 1991, 26). Quenching bronzes at a temperature of 586-798°C prevents the formation of a brittle δ phase. It has been reported that the Chinese high-tin bronze used for the production of the musical instrument *gong* is ductile and malleable and can be worked easily when quenched, but it is hard, springy and brittle when cooled slowly in air (Biot 1816, 185 cited in Goodway and Conklin 1987, 24). To investigate the workability of bronzes cast in this series, cold-working experiments have been conducted, the results of which will be discussed in chapter 7.4.

Table 4.4 ***Results of compositional analysis (wt%) by SEM/EDS of two water-quenched bronzes***

Sample no.	Nominal composition			ICP results			Analysed area	SEM/EDS results		
	Cu	Sn	Pb	Cu	Sn	Pb		Cu	Sn	Pb
125	77.0	23.0	0.0	76.5	23.5	n.d.	Matrix β	76.3	23.7	n.d.
							α	86.9	13.1	n.d.
							δ	68.8	31.2	n.d.
							Grain boundary	77.0	23.0	n.d.
128	96.0	2.0	2.0	96.3	1.8	1.9	Matrix	96.6	1.7	1.7
							α	98.6	1.0	0.4
							Grain boundary without Pb	91.0	8.4	0.6
							Pb on grain boundary	8.4	—	91.6

Table 4.5 ***Grain size and dendritic arm spacing of the bronzes cast in clay moulds***

Composition (wt%)			Dendritic arm spacing (μm)			Grain size (μm)		
Cu	Sn	Pb	Preheated, air-cooled	Unpreheated, air-cooled	Unpreheated, water-quenched	Preheated, air-cooled	Unpreheated, air-cooled	Unpreheated, water-quenched
98	2	0	—	—	—	143	132	116
94	6	0	63	61	36	975	833	2,900
90	10	0	69	58	39	1,167	486	638
85	15	0	67	52	45	1,560	1,140	700
77	23	0	—	—	—	—	—	714
96	2	2	53	61	58	3,800	890	827
92	2	6	—	41	32	1,271	740	780
88	2	10	—	—	—	167	160	109
92	6	2	57	51	34	855	394	488
88	6	6	52	40	28	944	1,300	809
84	6	10	55	52	33	613	840	850
88	10	2	78	79	31	1,100	1,467	353

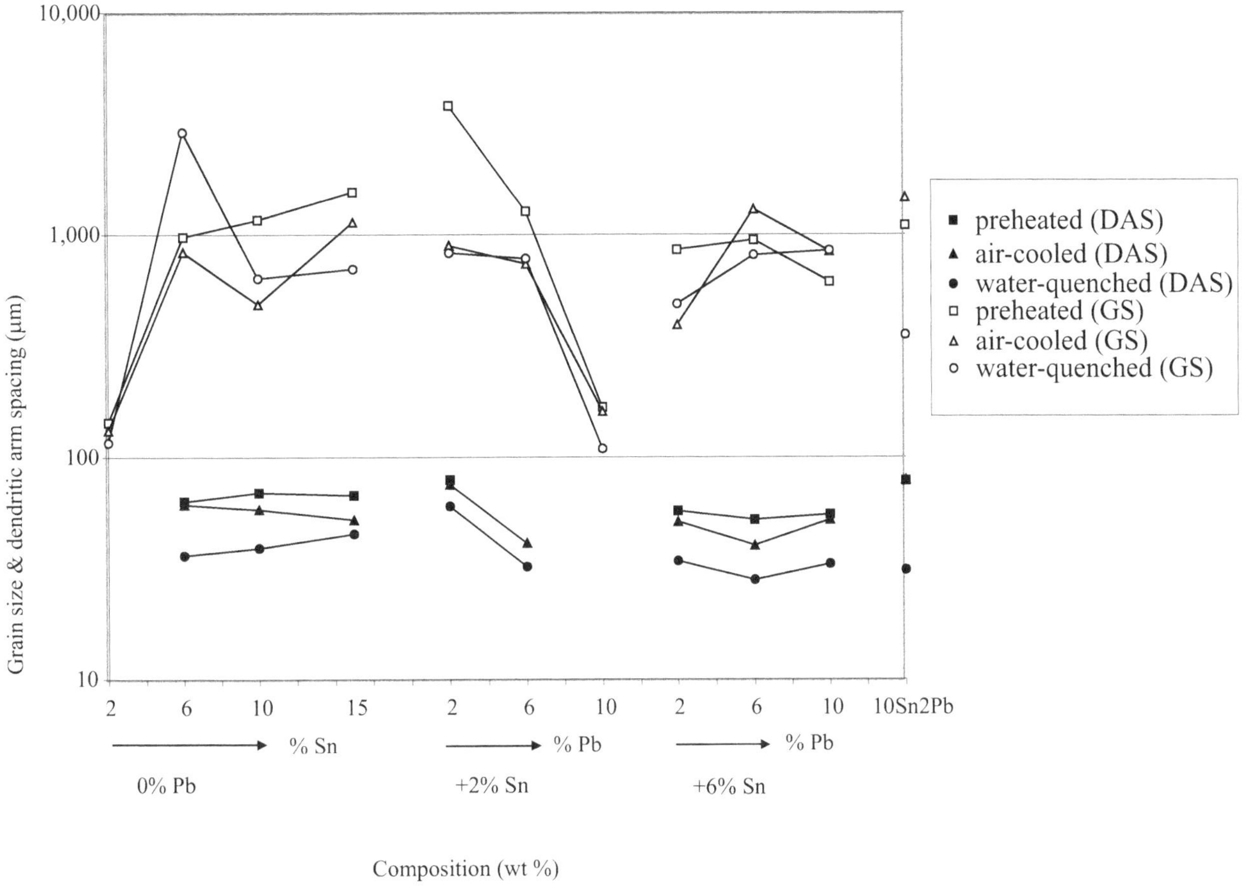

Figure 4.29 Grain sizes (GS) and dendritic arm spacing (DAS) of the bronzes cast in clay moulds

Table 4.6 ***Microhardness of bronzes of different composition cast in clay moulds***

Composition (wt%)			Preheated, air-cooled		Unpreheated, air-cooled		Unpreheated, water-quenched	
Cu	**Sn**	**Pb**	**Hv**	**SD**	**Hv**	**SD**	**Hv**	**SD**
98	2	0	58.5	2.7	60.4	5.2	71.1	2.5
94	6	0	77.1	8.7	77.6	7.1	92.2	8.6
90	10	0	99.8	10.9	88.9	3.5	108.1	12.8
85	15	0	109.2	11.0	122.3	10.7	140.0	23.9
77	23	0	225.4	44.4	247.8	33.7	250.0	26.0
96	2	2	60.7	1.8	54.6	4.7	66.2	4.0
92	2	6	52.1	4.6	56.2	5.9	59.1	4.3
88	2	10	51.9	10.5	59.5	3.5	60.9	4.7
92	6	2	81.7	5.9	74.4	10.5	107.1	20.9
88	6	6	74.6	1.9	70.7	10.0	82.3	9.5
84	6	10	72.1	7.9	75.2	7.3	74.4	11.3
88	10	2	101.2	6.6	96.7	14.7	133.4	19.0

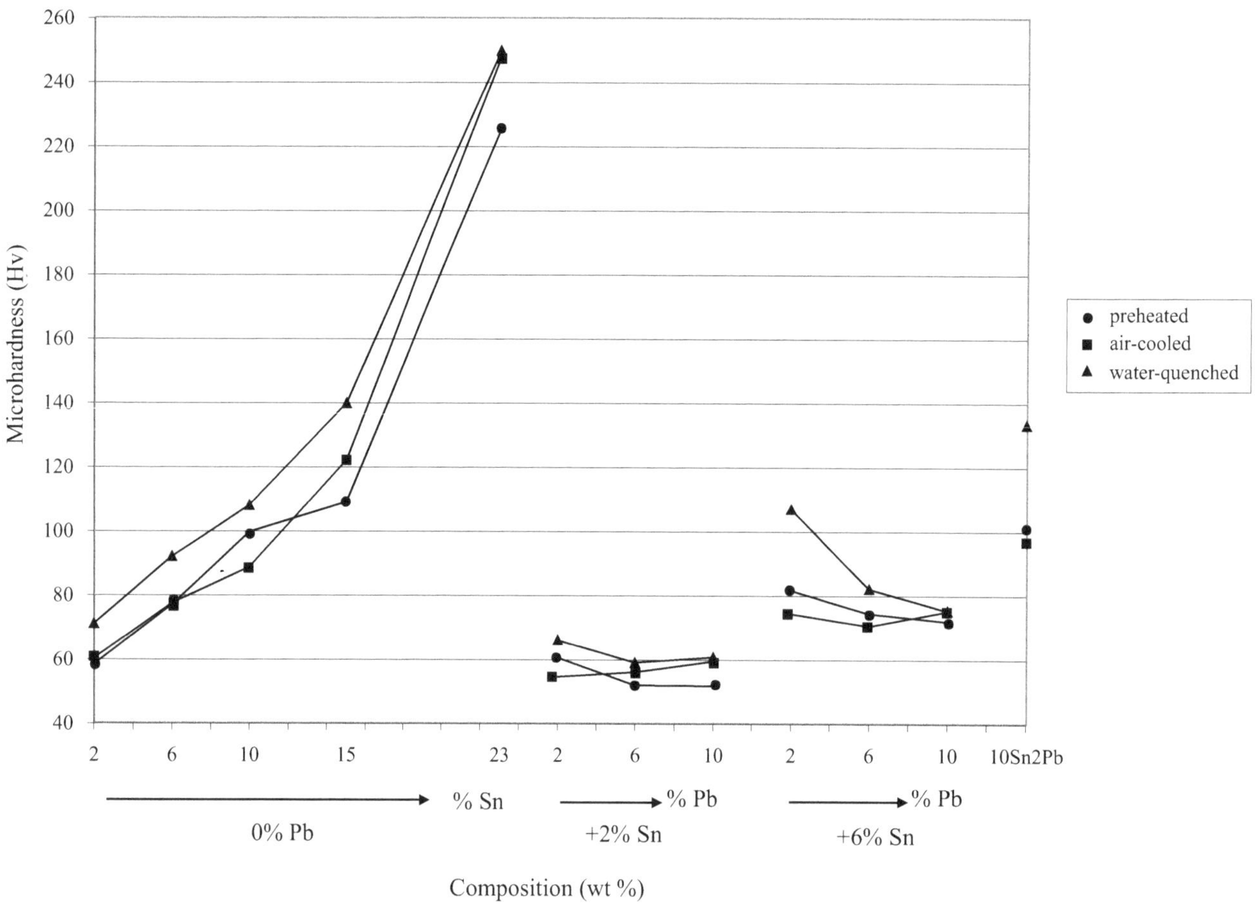

Figure 4.30 Vickers microhardness of the bronzes cast in clay moulds

4.5 Conclusions

Thirty-six bronze castings of twelve alloys, with three cooling rates for each alloy, have been made under controlled conditions in clay moulds. As was reported for the castings in sand moulds (chapter 3.2), great care has to be taken in the preparation of moulds, the melting of the metal and pouring the metal into the moulds, to ensure high-quality casts in clay moulds. It was found that objects cast in clay moulds have a finer surface than those cast in sand moulds; less work is therefore needed to finish the surface of the artefacts. However, some flashing seems to be unavoidable with clay moulds. The moulds could not be reused owing to damage caused during the release of the cast.

The experiments have also shown that preheating a fired clay mould is unnecessary for casting and makes little difference to the microstructure or hardness of the cast metal. There was, as expected, an appreciable difference between the microstructure of air-cooled casts and that of water-quenched casts.

The microstructures of the casts are complex. The low-tin (2% Sn) bronzes have only granular structures. However, the microstructures of the leaded low-tin bronzes vary (table 4.3): some have a granular structure, others a dendritic structure. A δ phase was seen in the water-quenched bronzes that contained 2% tin, much lower than is suggested by published equilibrium phase diagrams. The δ phase in the water-quenched casts looks darker in colour than that in the air-cooled bronzes. The β phase began to appear in water-quenched cast bronzes with 10% tin in the experiments, which is also much lower than has been shown in equilibrium phase diagrams.

A clear correlation between cooling rate and dendritic arm spacing was found, whereas grain size does not show such a clear correlation although there was a general trend towards larger grains in preheated and smaller grains in water-quenched casts, as was expected.

Microhardness values were dependent mainly on the composition of the metals, although almost all water-quenched casts in the series were found to have higher values than those from air-cooled moulds, whether preheated or not. It was interesting to note that, while the addition of lead generally leads to a decrease in hardness, the addition of only 2% lead seems to increase hardness in water-quenched casts.

5 Casting and metallography of bronzes cast in bronze moulds

5.1 Introduction

Bronze moulds appeared in the archaeological record from the Middle Bronze Age (Tylecote 1987, 221). However, there has been some debate about the actual use of bronze moulds. Childe (1930, 36) suggested that bronze moulds were probably not used for direct casting but for forming wax models, a suggestion taken up later by Tylecote (1986, 93), who pointed out that several bronze moulds had been found with the remains of lead axes in them and had therefore been used for making lead or wax patterns. Underwood (1958, 17) claimed that bronze moulds were 'clearly inferior for bronze casting', leading Coghlan (1968, 73) to remark that 'no firm conclusion had been reached as to the precise purpose of the bronze moulds'.

Stansby (1984, 46), on the other hand, maintained that it is entirely feasible to use metal moulds for the direct casting of bronze. Tylecote accepted that bronze moulds were used in this way in his later book (1987, 223): he ascertained that most of the bronze moulds found in the British Isles were used for direct casting of bronze implements. Stansby (1984, 46) pointed out that there is no evidence to suggest that bronze moulds should have anything but a long life. Tylecote (1986, 92) questioned this on the grounds that if they were used for direct casting they 'would need to be replaced every 50 or so castings'. However, unpublished experiments carried out in the 1950s in which bronze implements were cast in bronze moulds revealed that one of the moulds did not show any ill effects after being used for a run of 15 castings (Drescher, cited in Leahy 1977, 2; Mohen 1978, 28).

It is against the background of these conflicting arguments that experiments with bronze moulds were carried out in Sheffield, first by Swiss (Swiss and Ottaway in press) and later in the series of experiments reported here.

The composition of a bronze mould has to be decided prior to casting, careful consideration being given to the issue of fusion between the molten metal being cast and the metal of the mould. This concern was raised as early as 1913 by Gotze (cited in Mohen 1978, 27), who suggested that molten bronze would stick to the inside of a bronze mould. Tylecote (1987, 221) thought that bronze moulds were usually made of the same bronze as the castings, but Stansby's analyses (1984, 12) demonstrated that bronze moulds were typically made from an alloy containing 12% tin and between 0 and 2% lead. Compositional analysis of two bronze socketed axe moulds from Heathery Burn near Worthing showed that they contained 12.5 and 15% tin, respectively (Leahy 1977, 4). As the experiments reported here involved the casting of bronzes of a whole range of compositions, all of which the bronze moulds would have to accommodate, it was decided that an alloy consisting of 90% copper and 10% tin would be most suitable for the bronze moulds.

To reduce the danger of fusion between the mould and the liquid metal and to ease the separation of the cast from the mould, soot is often used to dress the mould (Tylecote 1976, 33; 1987, 221). Voce (1975, 137) dressed the internal surfaces of the mould with graphite paste. Swiss dressed his bronze mould with soot for one casting, but carried out another casting without dressing. The latter did not prove detrimental to the casting; no fusion between mould and hot metal was observed and the axe was easily removed (Swiss and Ottaway in press).

To compare the effect of different mould dressings on the quality of the cast, two preliminary castings were made: the inside of one mould was rubbed with charcoal and the other was coated with powdered graphite. Both moulds were flame-warmed and air-cooled. As there was little difference in the quality of cast, charcoal was used for coating the voids in the casting series reported here since it would have been readily available in prehistory.

5.2 Mould-making

A pattern of the mould made of resin is shown in figure 5.1. To facilitate comparison, the mould was initially designed to be as similar as possible in size and the way the feeder, or sprue, was positioned to the clay moulds used in the series of experiments.

Two bronze moulds were cast in sand moulds. The details of the procedure were similar to those described in chapter 3.2 with the following changes:

- The drag box was positioned on the working table with the two halves of the resin pattern suitably placed and dusted with parting powder (fig. 5.1). The drag box was made as outlined in chapter 3.2 and the completed drag box is shown in figure 5.2.
- The cope box was then made as outlined in chapter 3.2. The pattern was taken out. A runner was made at one side and a riser at the other side of the cavity between the two halves of the pattern. Several vents were made. The completed drag and cope boxes are shown in figure 5.3.
- The two halves of the cast bronze mould, with prominent flashing, are shown in figure 5.4, the finished mould halves with the flashing removed in figure 5.5. The assembled bronze mould is shown in figure 5.6.
- A runner bush of sand was fitted on top of the mould to guide the metal while it was being poured (fig. 5.7) and the two halves of the mould were fastened together with G-clamps (fig. 5.8).

This bronze mould, with the feeder on the top part of the mould, was used for a trial casting. The void of the mould was rubbed with charcoal as a mould dressing to ease the removal of the cast axe. A bronze with 2% tin was used and the metal was heated to 1260°C. No problems with pouring (fig. 5.8) or removing the axe from the mould were experienced in this part of the experiment.

The only problem was that the runner bush was bigger than the feeder so that it was difficult to remove the axe from the top part of the mould (fig. 5.9). After removal of the axe, the mould remained undamaged and could have been used again (fig. 5.10). However, it was considered too time-consuming and labour-intensive to continue using this type of runner bush and feeder.

To prevent the runner bush and feeder from sticking to the top part of the mould, a wedge-shaped feeder was made and a 10% tin bronze was cast in this mould. The result was very disappointing. The bottom part of the mould, where the molten metal had hit it, had melted. This made removal of the axe without damaging the mould impossible. It is possible that the hot spot was not dressed well enough with charcoal. In any case the problem was solved by cutting a funnel-shaped feeder into the side of the mould (figs. 5.11-5.12). This made the removal of the axe much easier after casting. Two moulds of this kind were made and used for the experimental series.

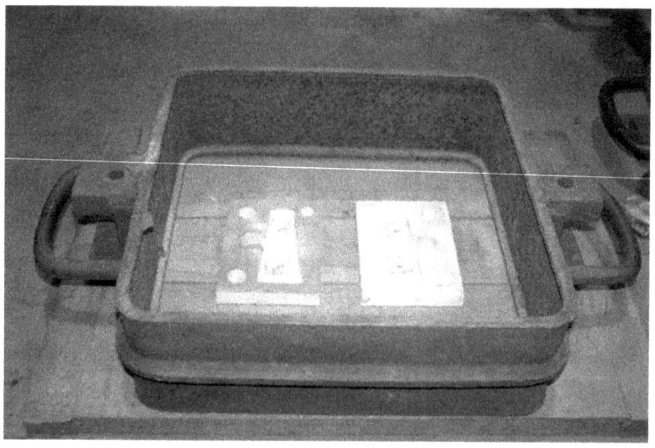

Figure 5.1 Drag box with the pattern

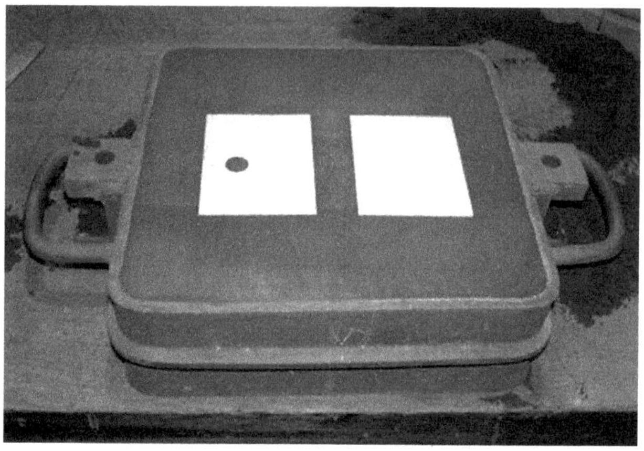

Figure 5.2 Completed drag box

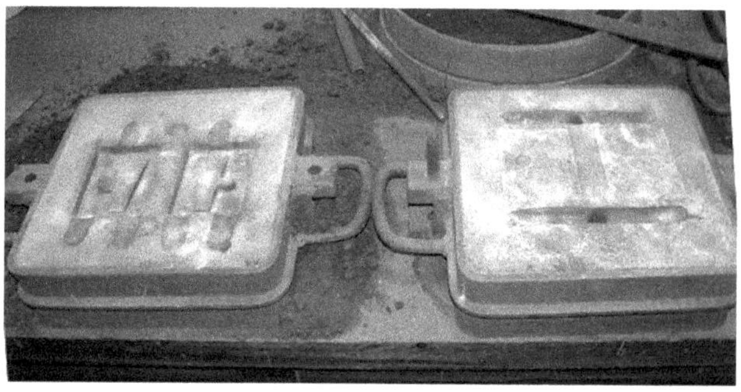

Figure 5.3 Completed drag and cope boxes for the bronze mould

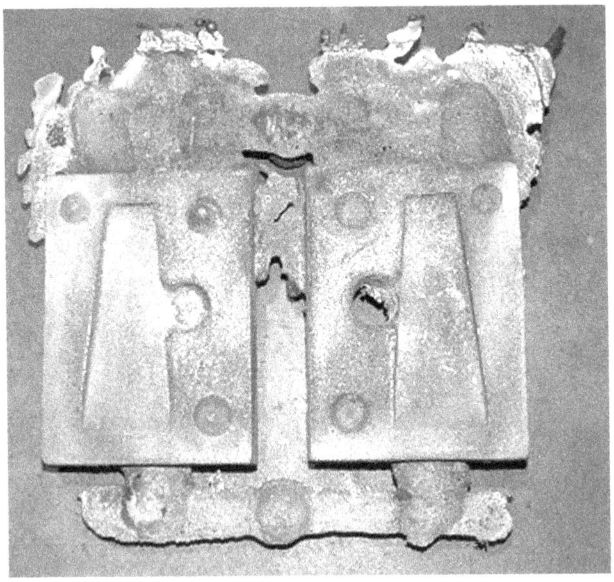

Figure 5.4 The two halves of the first bronze mould, with flashing

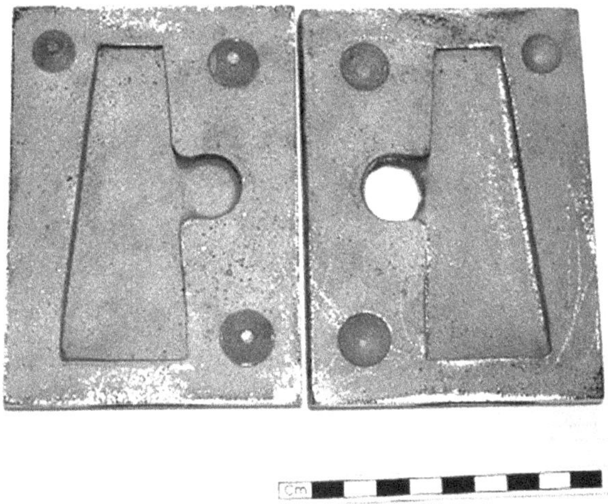

Figure 5.5 Finished halves of the first bronze mould

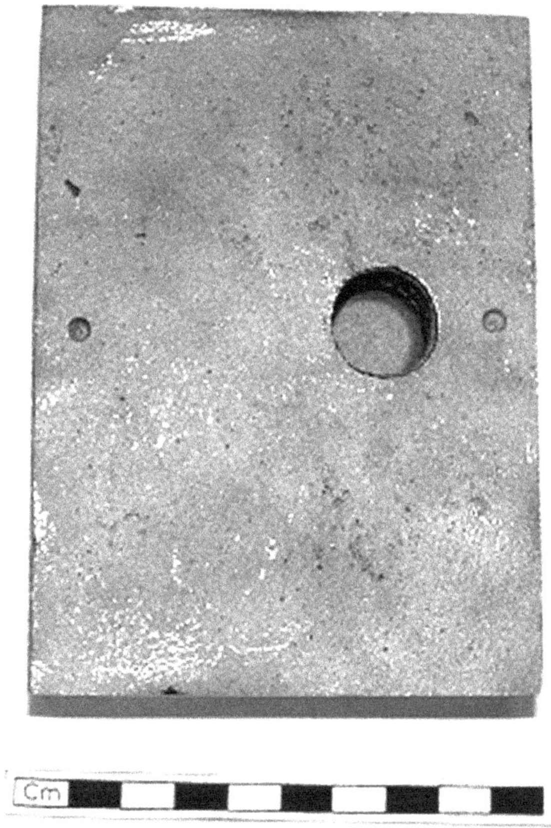

Figure 5.6 The first bronze mould, viewed from top

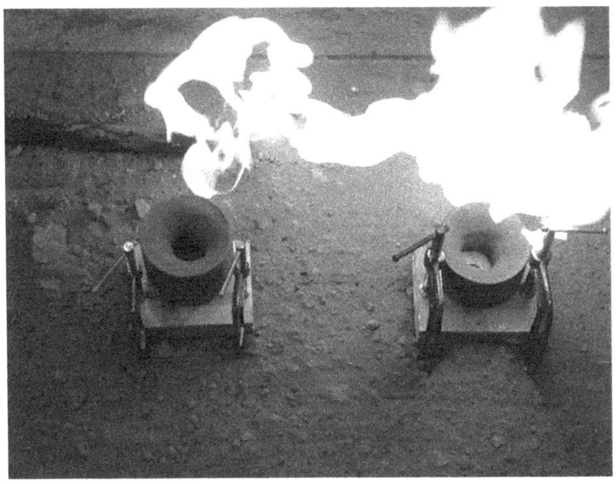

Figure 5.7 Flame-warmed bronze moulds ready for casting

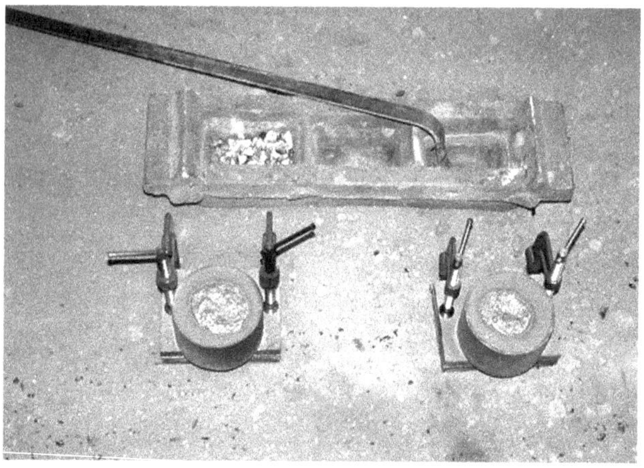

Figure 5.8 The clamped moulds after casting with a reservoir of metal filling the runner bush

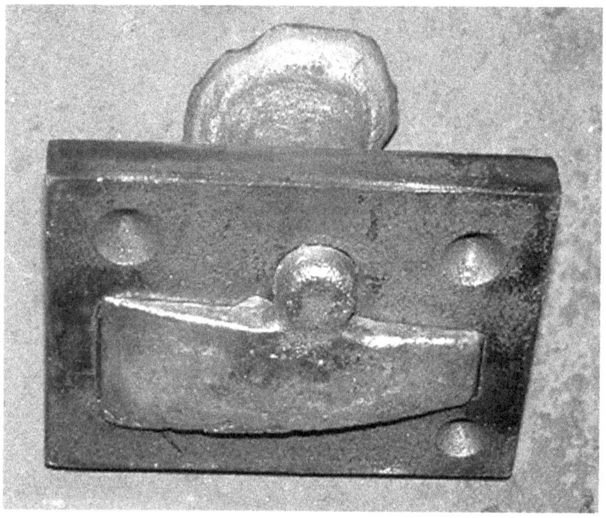

Figure 5.9 Solidified feeder and runner bush attached to the upper part of the first mould

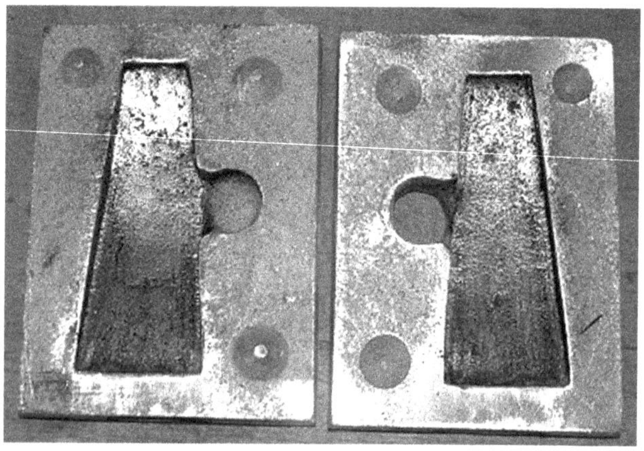

Figure 5.10 The first mould after one use

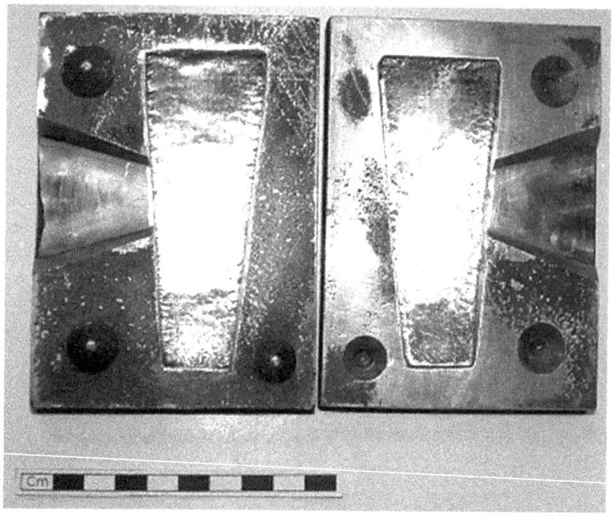

Figure 5.11 Funnel-shaped feeders cut at the sides of the bronze mould halves

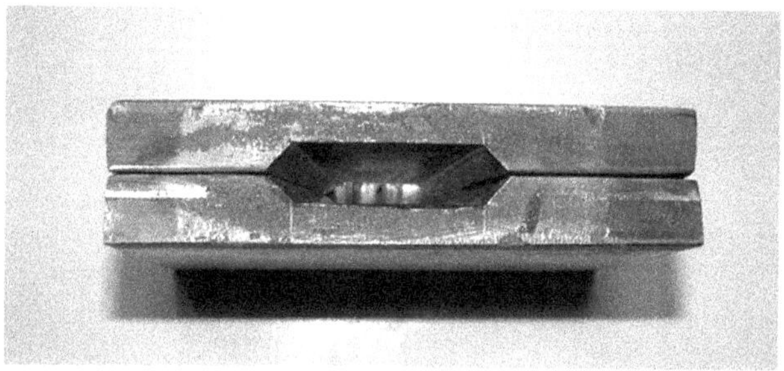

Figure 5.12 Side view of the second mould with the feeder at the side

5.3 Experimental casting

In the initial experiments, the cavity of one mould was smoothed with files and sanded with emery paper, and the other one was just sand-blasted. As expected, the axe cast in the smoothed mould had a better surface than the other one. Subsequently, therefore, the cavities of the moulds were always smoothed and sanded.

The optimum conditions for making bronze moulds having thus been established and two bronze moulds produced in this way, the following experimental series was carried out: 12 different alloys and two cooling regimes, i.e. casting into preheated (to 350°C) and into flame-warmed moulds (estimated to be lower than 350ºC), were used in the casting experiments. For safety reasons, water-quenching was abandoned in the series of bronze mould castings.

The melting of the metal and control of temperature were as described in chapter 2.2. The results of the castings are presented here.

5.4 Results and discussion

The experiments showed that the way in which the moulds were preheated prior to casting affected their longevity. In the flame-warmed mould a small crack appeared in the middle of the mould during the sixth use and widened during the seventh casting. The preheated mould did not show any sign of damage up to the ninth casting. Both moulds were damaged so severely that they could not be used again after the tenth casting (fig. 5.13). From this point it was also very difficult to remove the axe from the mould.

The series of experiments showed that bronzes cast in bronze moulds had a better surface appearance than those cast in clay or sand moulds and there was very little flashing (fig. 5.14). The axes could thus be finished more speedily (fig. 5.15a and b) than those cast in sand or clay moulds.

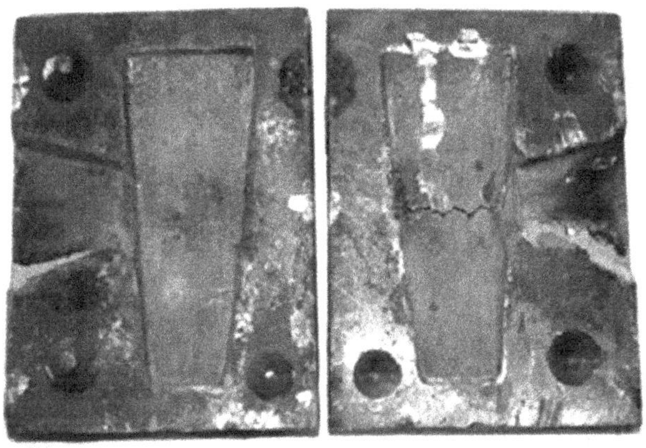

Figure 5.13 The damaged mould after it had been used for ten castings

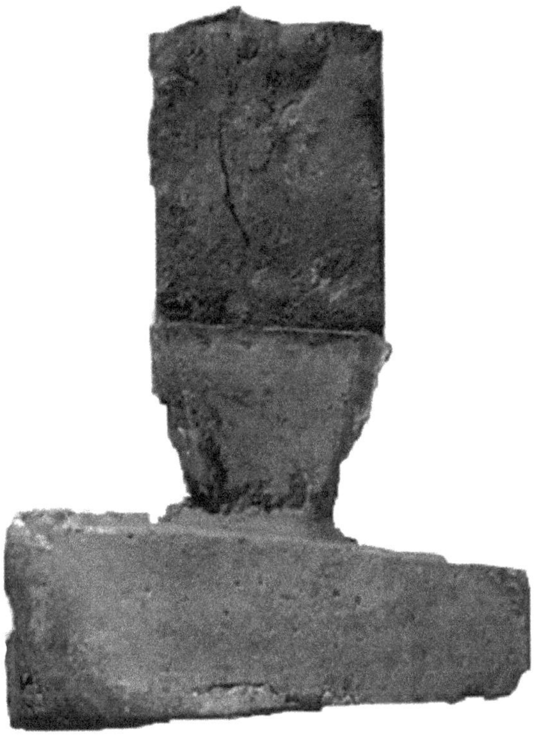

Figure 5.14 An axe with feeder cast in a bronze mould

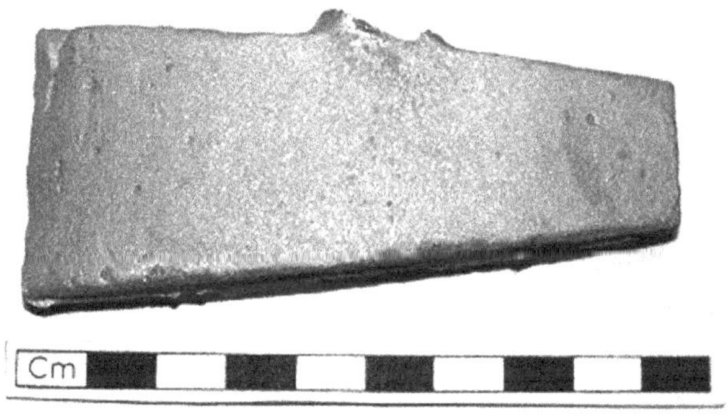

a. Top view

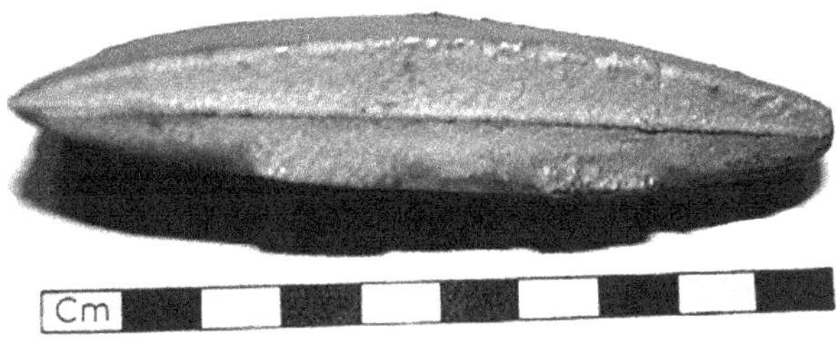

b. Side view, showing casting seam

Figure 5.15 An axe cast in a bronze mould after sand-blasting

5.4.1 Microstructure of the casts

The cast axes were cut in half and micropores were found to be present in most of the sections, as had been observed for casts made in clay and sand moulds. A selection of photomicrographs is shown in black and white in figures 5.16-5.22. These figures are shown again in colour in appendix 1 on the CD accompanying this volume. A complete series of photomicrographs of bronzes cast in bronze moulds is given in appendix 4 on the CD. All results are summarised in tabular form in table 5.1. Details are presented and discussed below.

Metallographic examination has shown that all bronzes cast in bronze moulds have a dendritic structure. In most cases there is little difference in the microstructure of bronzes cast in the preheated moulds and those cast in the flame-warmed moulds. This could be due to the similarity in mould temperature after preheating prior to casting. However, as discussed earlier the moulds had to be warmed up for safety reasons.

The dendrites are coarser in some bronzes than in others, as can be seen in the 2% tin bronze, which has coarse dendrites (fig. 5.16), compared to the 10% tin bronze, where the dendrites are finer but longer (fig. 5.17). Both bronzes had been cast in moulds which had been preheated to 350°C. This effect is most probably due to the different composition of the metal. A transition from columnar dendritic structure at the surface to granular structure at the centre can be seen in bronzes with 2% tin cast in a flame-warmed mould (fig. 5.18a and b). Heterogeneity in the size of dendrites can be present in a single bronze, as is illustrated in figure 5.19, a 2% tin and 2% lead bronze cast in a flame-warmed mould, where both coarse and fine dendrites can be seen. This phenomenon is present in 9 out of 24 bronzes cast in bronze moulds in this experimental series. Coarse but very short dendrites are also seen in some of the casts, such as the 2% tin, 10% lead bronze, which was cast in a mould preheated to 350°C (fig. 5.20). There is no correlation between this phenomenon and the composition of the bronzes. The δ phase is present in bronzes containing 6% tin or more.

Despite little apparent difference between the bronzes cooled according to the two different regimes, the bronzes cast in flame-warmed moulds sometimes had a more pronounced dendritic structure and more $\alpha+\delta$ eutectoids than those cast in preheated moulds, as can be seen in the 10% tin bronzes (fig. 5.21a and b) and the 6% tin and 6% lead bronzes (fig. 5.22a and b). This might be explained by the fact that in all the experiments the molten metal was first poured into the preheated and then into the flame-warmed mould and was therefore very slightly cooler in the latter. The contrast between the two cooling rates is comparable to the air-cooling versus water-quenching used for the sand and clay mould casting.

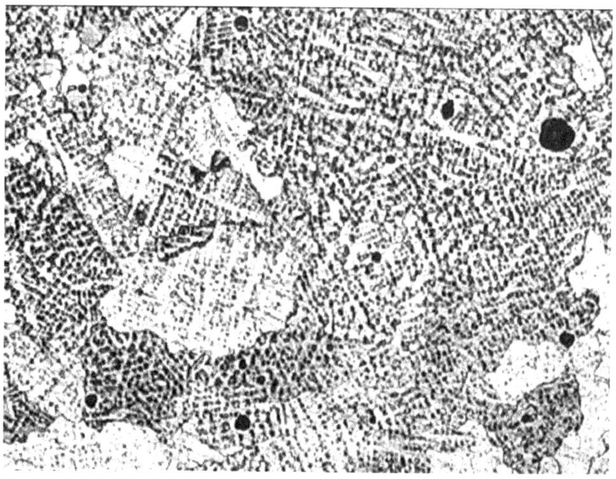

Figure 5.16 Photomicrograph of 2% Sn bronze, mould preheated to 350°C, showing a dendritic structure (image width 1.3 mm)

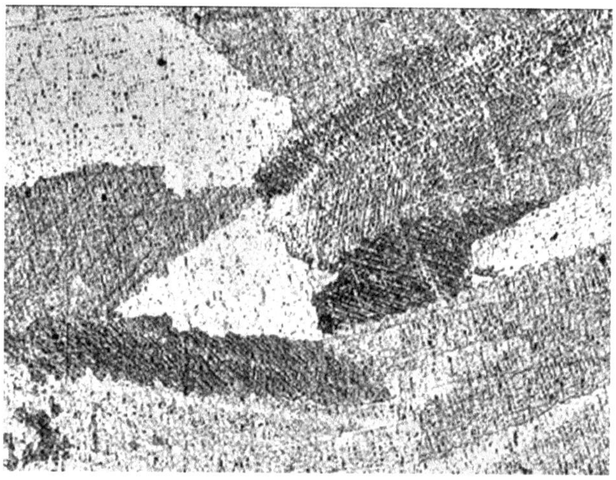

Figure 5.17 Photomicrograph of 10% Sn bronze, mould preheated to 350°C, showing a dendritic structure. The dendrites are finer but longer than those in figure 5.16 (image width 1.3 mm)

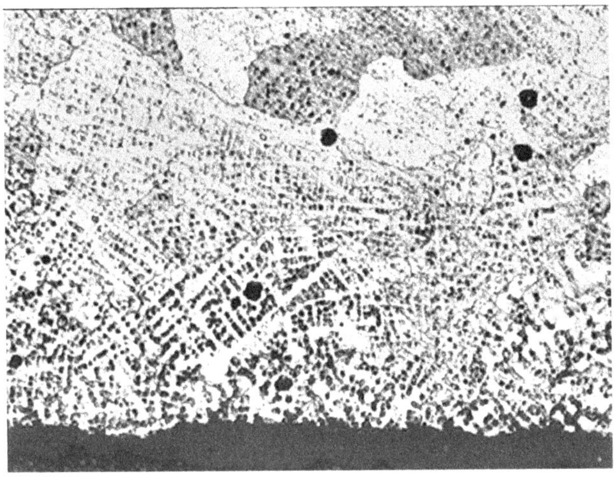

a. Dendritic structure on the surface

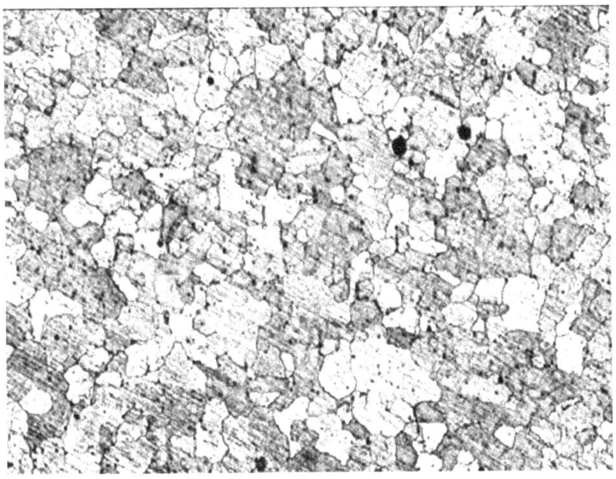

b. Granular structure in the centre

Figure 5.18 Photomicrographs of 2% Sn bronze, mould warmed by flame (image width 1.3 mm)

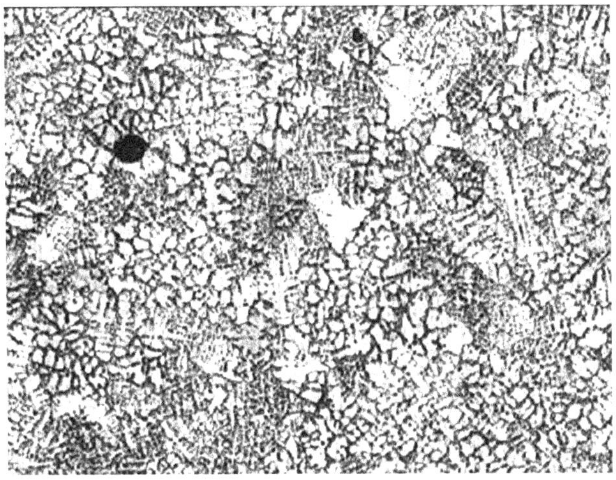

Figure 5.19 Photomicrograph of bronze with 2% Sn + 2% Pb, mould warmed by flame, showing both coarse and fine dendrites (image width 1.3 mm)

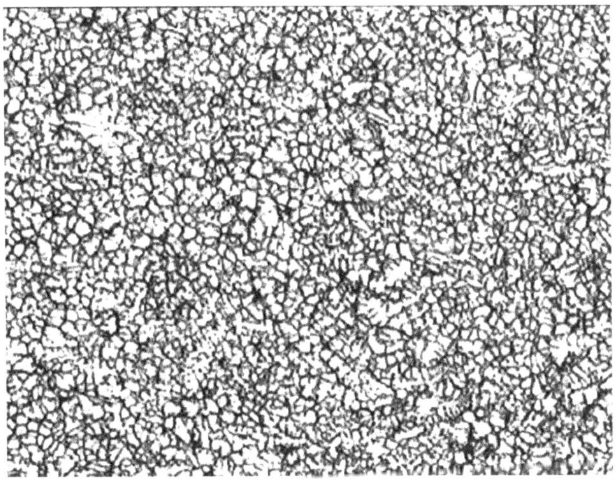

Figure 5.20 Photomicrograph of bronze with 2% Sn + 10% Pb, mould preheated to 350°C, showing coarse but short dendrites (image width 1.3 mm)

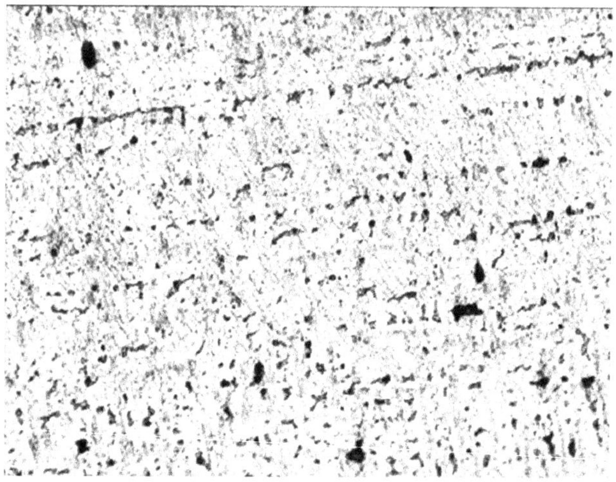

a. Mould preheated to 350°C

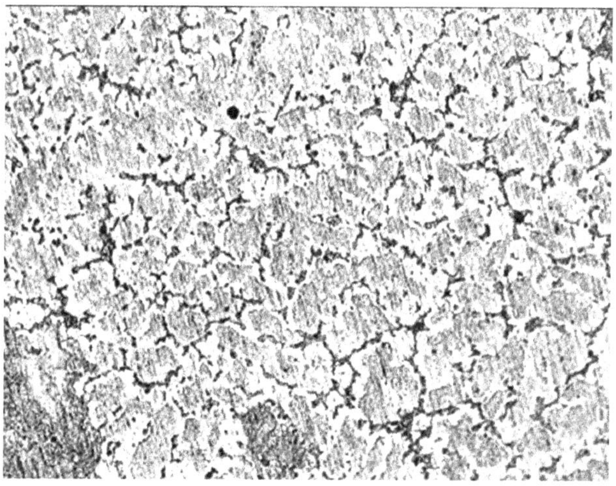

b. Mould flame-warmed

Figure 5.21 Photomicrographs of 10% Sn bronzes, showing more pronounced $\alpha+\delta$ eutectoids in the bronze cast in the flame-warmed mould (b) than in that cast in the preheated mould (a) (image width 0.33mm)

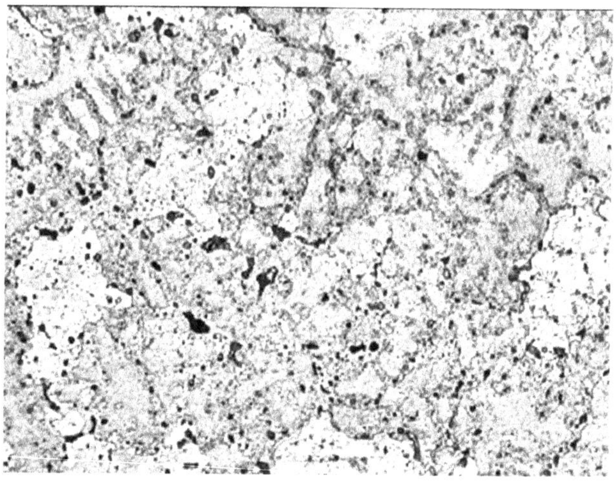

a. Mould preheated to 350°C

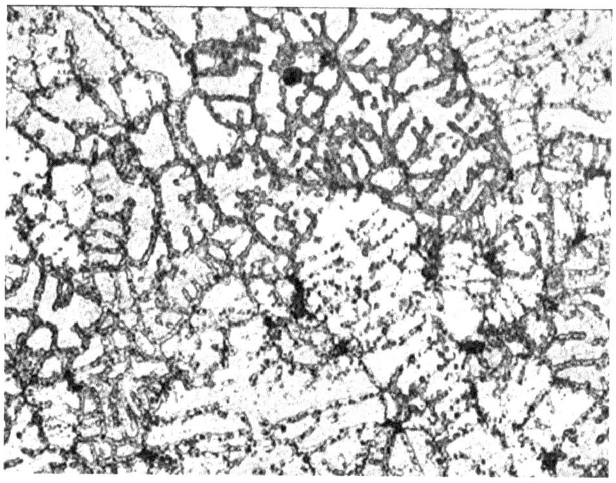

b. Mould flame-warmed

Figure 5.22 Photomicrographs of bronzes with 6% Sn+6% Pb, showing more pronounced dendritic structure and more $\alpha+\delta$ eutectoids in the bronze cast in the flame-warmed mould (b) than in that cast in the preheated mould (a) (image width 0.33 mm)

Table 5.1 ***Composition, cooling regimes and microstructures of the bronze axes cast in the bronze moulds***

Sample no.	Composition[1] (wt %)		Cooling method	Microstructure	Figure no.
	Sn	Pb			
203	2	—	Preheated, air-cooled	Dendritic without α+δ eutectoids	5.16
204	2	—	Flame-warmed, air-cooled	Dendritic at the surface and granular in the centre without α+δ eutectoids	5.18
205	6	—	Preheated, air-cooled	Dendritic with α+δ eutectoids on grain boundaries	—
206	6	—	Flame-warmed, air-cooled	Dendritic with α+δ eutectoids on grain boundaries	—
207	10	—	Preheated, air-cooled	Dendritic with many α+δ eutectoids	5.17, 5.21a
208	10	—	Flame-warmed, air-cooled	Dendritic with many α+δ eutectoids	5.21b
209	15	—	Preheated, air-cooled	Dendritic with many α+δ eutectoids	—
210	15	—	Flame-warmed, air-cooled	Dendritic with many α+δ eutectoids	—
211	23	—	Preheated, air-cooled	Smaller grain size than the flame-warmed sample, many α+δ eutectoids	—
212	23	—	Flame-warmed, air-cooled	Very long dendrites with many α+δ eutectoids	—
213	2	2	Preheated, air-cooled	Dendritic without α+δ eutectoids	—
214	2	2	Flame-warmed, air-cooled	Dendritic without α+δ eutectoids, uneven dendritic arm spacing	5.19
215	2	6	Preheated, air-cooled	Dendritic without α+δ eutectoids; short dendrites	—
216	2	6	Flame-warmed, air-cooled	Dendritic without α+δ eutectoids	—
217	2	10	Preheated, air-cooled	Dendritic without α+δ eutectoids; short dendrites	5.20
218	2	10	Flame-warmed, air-cooled	Dendritic without α+δ eutectoids	—
219	6	2	Preheated, air-cooled	Dendritic without α+δ eutectoids	—
220	6	2	Flame-warmed, air-cooled	Dendritic without α+δ eutectoids	—
221	6	6	Preheated, air-cooled	Dendritic with a few α+δ eutectoids; short dendrites	5.22a
222	6	6	Flame-warmed, air-cooled	Dendritic with more α+δ eutectoids than the preheated sample; short dendrites	5.22b
223	6	10	Preheated, air-cooled	Dendritic with a few α+δ eutectoids	—
224	6	10	Flame-warmed, air-cooled	Dendritic with a few α+δ eutectoids	—
225	10	2	Preheated, air-cooled	Dendritic with many α+δ eutectoids; long dendrites	—
226	10	2	Flame-warmed, air-cooled	Dendritic with more α+δ eutectoids than the preheated sample; long dendrites	—

Note: [1] The balance is copper.

5.4.2 Grain size and dendritic arm spacing

Grain size and dendritic arm spacing were measured as described in chapter 2.3. The results of the measurements are listed in table 5.2 and plotted in figure 5.23. The figure shows that grain size, and particularly dendritic arm spacing, both give similar values for the two cooling regimes and do not show a clear correlation with cooling rates.

Table 5.2 ***Grain size and dendritic arm spacing of the bronzes cast in bronze moulds***

Composition (wt%)			Dendritic arm spacing (μm)		Grain size (μm)	
Cu	**Sn**	**Pb**	**preheated, air-cooled**	**flame-warmed, air-cooled**	**preheated, air-cooled**	**flame-warmed, air-cooled**
98	2	0	14	15	143	250
94	6	0	12	13	143	110
90	10	0	12	13	330	500
85	15	0	13	13	500	170
77	23	0	—	—	500	250
96	2	2	18	14	180	150
92	2	6	11	13	73	110
88	2	10	16	12	35	110
92	6	2	14	11	250	250
88	6	6	12	15	170	75
84	6	10	9	11	200	100
88	10	2	14	12	330	330

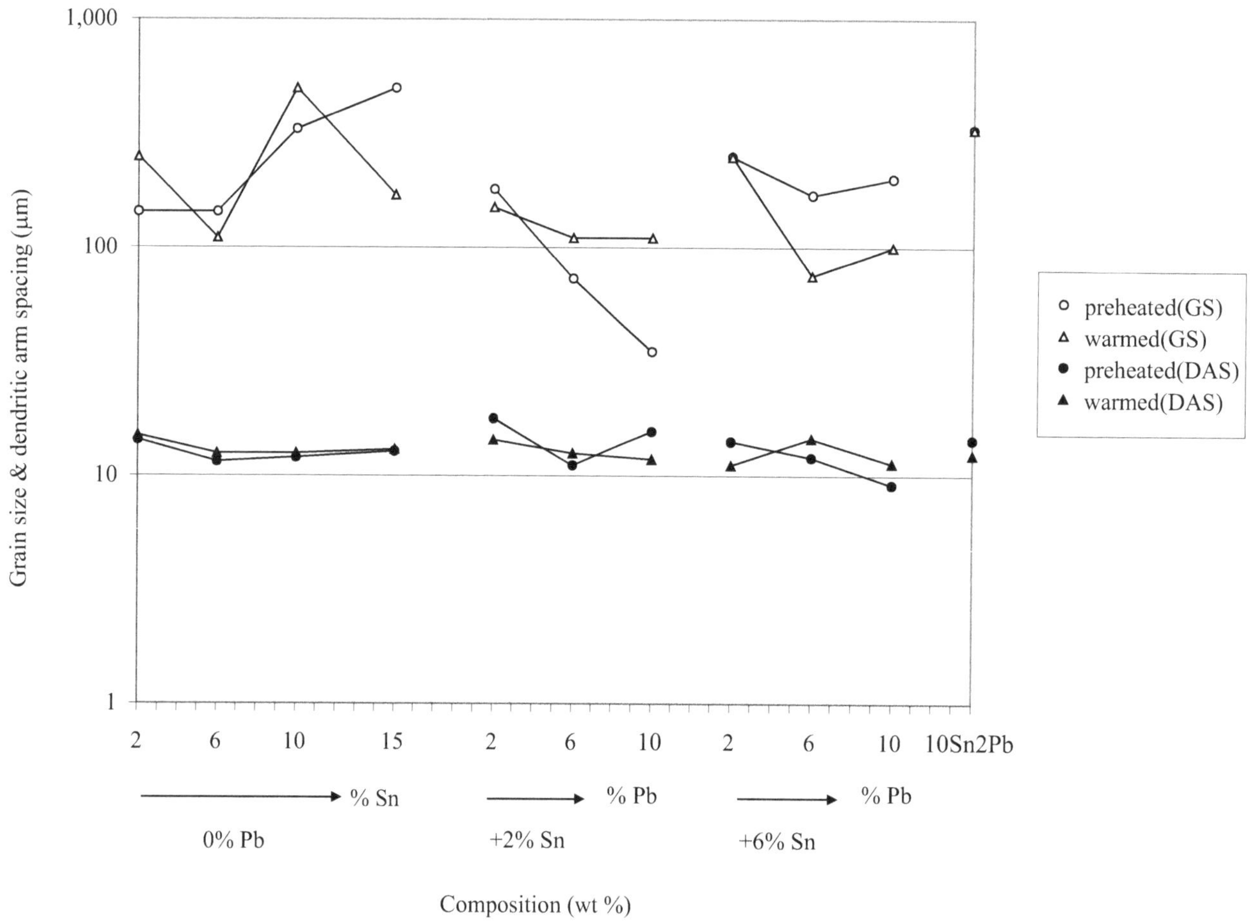

Figure 5.23 Grain sizes and dendritic arm spacing of the bronzes cast in bronze moulds

5.4.3 Microhardness

The microhardness was measured as described in chapter 2.3. The results are listed in table 5.3. The correlation of microhardness of the various compositions with different cooling rates is shown in figure 5.24. The graphs, similar to the plot of grain size and dendritic arm spacing described above, show little difference in hardness values between the two different cooling regimes. As expected, the hardness of tin bronzes increases with increasing content of tin, whereas the addition of lead to tin bronzes tends to reduce the hardness slightly.

Table 5.3 ***Microhardness of bronzes cast in bronze moulds***

Composition (wt %)			Preheated		Flame-warmed	
Cu	Sn	Pb	Hv	SD	Hv	SD
98	2	0	63.1	5.0	57.9	7.4
94	6	0	78.8	9.2	82.0	4.1
90	10	0	117.6	10.1	112.2	12.7
85	15	0	138.0	6.7	159.4	7.0
77	23	0	247.6	20.1	238.6	14.3
96	2	2	58.8	1.9	70.0	9.2
92	2	6	58.0	3.9	59.6	5.4
88	2	10	59.7	6.5	55.9	0.4
92	6	2	76.3	2.0	82.6	1.5
88	6	6	77.6	2.7	85.6	4.9
84	6	10	73.7	4.1	79.3	4.2
88	10	2	96.3	6.2	108.4	13.4

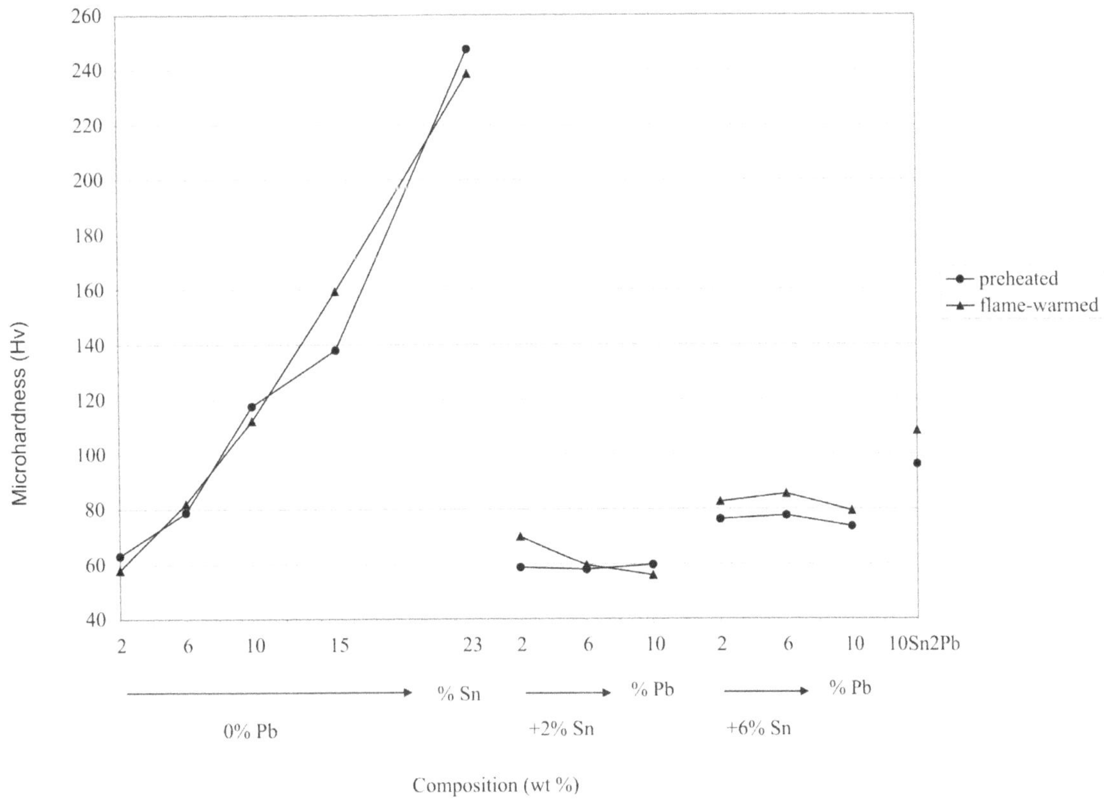

Figure 5.24 Vickers microhardness of the bronzes cast in bronze moulds

5.5 Conclusions

Twenty-four bronze castings, of 12 alloys and with two cooling rates for each alloy, have been carried out under controlled conditions using bronze moulds. The results have shown that the preheated mould lasted longer than the flame-warmed one. The former was used nine times and the latter six times before any damage appeared.

The bronzes cast in the bronze moulds had almost no flashing and the surfaces of the casts were quite smooth. All bronzes cast in bronze moulds have a dendritic structure and only the core of the 2% tin bronze had any granular structure. Little difference in microstructure between bronzes cast in preheated moulds and those cast in flame-warmed moulds could be observed. There is no clear correlation between cooling rates and dendritic arm spacing, nor is there a correlation between cooling rates and microhardness values, which could be due to the lack of significant difference in temperature in the two moulds prior to casting.

6 Comparison of the microstructure of bronzes cast in three different moulding materials

6.1 Introduction

Twelve alloys were used for the experimental castings in moulds of sand, clay and bronze. The results of the casting experiments have been presented in chapters 3-5 for each moulding material. To compare the effects of the moulding materials on the properties of the casts, the casts studied in this chapter are mainly air-cooled ones as this cooling regime has been employed on all mould materials. Water-quenched bronzes cast in sand and clay moulds and bronzes cast in preheated clay and bronze moulds will also be discussed for comparison.

6.2 Microstructure and dendritic arm spacing

The microstructures of all casts made in moulds of the three different materials, all air-cooled, are summarised in table 6.1; water-quenched, or chill-cast, bronzes cast in sand and clay moulds are summarised in table 6.2 and bronzes cast in preheated clay and bronze moulds in table 6.3.

It is interesting to observe that the bronzes with 2% tin, cast in sand and clay moulds and air-cooled, have a granular structure, while one with the same composition cast in a bronze mould has a dendritic structure at the surface and granular structure in the centre (table 6.1). Among the leaded bronzes with 2% tin, those cast in sand moulds with lower lead (2 and 6%) have a partly granular and partly dendritic structure, whereas the higher (10%) leaded cast has a purely granular structure. Two of these bronzes (2 and 6% lead) cast in clay moulds have a dendritic structure whereas the 10% leaded cast has a granular structure. Leaded bronzes, including those with varying amounts of tin, cast in bronze moulds, have dendritic structures.

There were no $\alpha+\delta$ eutectoids in any of the bronzes with 2% tin, irrespective of moulding material or lead content. All casts with 6% tin have a dendritic structure, irrespective of either moulding material or lead content. $\alpha+\delta$ eutectoids began to appear in all tin bronzes with 6% tin or higher, irrespective of the moulding material. However, they are absent from all leaded bronzes with 6% tin cast in sand moulds, and there are very few in bronzes of the same composition cast in clay moulds, though more are found in those cast in bronze moulds. All bronzes with 10% tin or more have dendritic structures with $\alpha+\delta$ eutectoids. The higher the tin content the more $\alpha+\delta$ eutectoids there are.

The differences in microstructure can also be illustrated by dendritic arm spacing (DAS). Figure 6.1 shows that the bronzes cast in the bronze moulds have the smallest DAS (9-18 μm) whereas those cast in the sand moulds have DAS values which are considerably larger (30-60 μm) and are partly overlapping with values of the bronzes cast in the clay moulds which have the largest DAS (40-80 μm).

The difference in the values of DAS is due to the difference in cooling rates. Staniaszek and Northover (1983, 263) show that bronzes cast in clay moulds have a slower cooling rate than those cast in stone and metal moulds (fig. 6.2). Eccleston and Ottaway (2002, 189 and fig. 4) also report that in their experimental casting bronzes cooled more slowly in sand moulds containing 30% clay than in those containing 20% clay.

In addition to moulding material, the cooling rate is also affected by the composition of the alloy, the casting temperature, the mould mass, the pretreatment of the mould, and the size of the object to be cast. In the experiments presented here the average DAS for bronzes cast in bronze moulds is 13.1 μm, compared to 42.1 μm in sand moulds and 56.6 μm in clay moulds. Staniaszek and Northover (1983, 263) report that in their experiments the average dendritic arm spacing for an alloy cast in a steel mould is 8.2 μm, while the figure is 25.6 μm for an alloy cast in a clay mould. Junk (2003) obtained DAS values for bronzes cast in sand moulds between 20 and 50 μm. Bearing in mind that it is difficult to compare dendritic arm spacing obtained by different researchers carrying out different casting experiments using different alloys, moulds and patterns, the experiments have nevertheless shown a clear trend: clay, followed by sand mould castings, with much lower cooling rate, have a much larger DAS than metal mould castings, which have faster cooling rates and smaller DAS.

Table 6.1 ***Microstructures of air-cooled axes cast in different moulding materials***

Composition (wt %)			Moulding material		
Cu	**Sn**	**Pb**	**Sand**	**Clay**	**Bronze**
98	2	—	Granular without $\alpha+\delta$ eutectoids	Granular without $\alpha+\delta$ eutectoids	Dendritic at the surface and granular in the centre without $\alpha+\delta$ eutectoids
94	6	—	Dendritic with very few $\alpha+\delta$ eutectoids	Dendritic with very few $\alpha+\delta$ eutectoids	Dendritic with $\alpha+\delta$ eutectoids on grain boundaries
90	10	—	Dendritic with $\alpha+\delta$ eutectoids	Dendritic with $\alpha+\delta$ eutectoids	Dendritic with many $\alpha+\delta$ eutectoids
85	15	—	Dendritic with many $\alpha+\delta$ eutectoids	Dendritic with many $\alpha+\delta$ eutectoids	Dendritic with many $\alpha+\delta$ eutectoids
77	23	—	Dendritic with many $\alpha+\delta$ eutectoids	Dendritic with massive $\alpha+\delta$ eutectoids	Very long dendrites with massive $\alpha+\delta$ eutectoids
96	2	2	Both dendritic and granular without $\alpha+\delta$ eutectoids	Dendritic without $\alpha+\delta$ eutectoids	Dendritic without $\alpha+\delta$ eutectoids, uneven dendritic arm spacing
92	2	6	Both dendritic and granular without $\alpha+\delta$ eutectoids	Dendritic without $\alpha+\delta$ eutectoids	Dendritic without $\alpha+\delta$ eutectoids
88	2	10	Granular without $\alpha+\delta$ eutectoids	Granular without $\alpha+\delta$ eutectoids; Pb on both the grain boundaries and within the grains	Dendritic without $\alpha+\delta$ eutectoids
92	6	2	Dendritic without $\alpha+\delta$ eutectoids	Dendritic with very few $\alpha+\delta$ eutectoids	Dendritic without $\alpha+\delta$ eutectoids
88	6	6	Dendritic without $\alpha+\delta$ eutectoids	Dendritic with very few $\alpha+\delta$ eutectoids	Dendritic with $\alpha+\delta$ eutectoids; short dendrites
84	6	10	Dendritic without $\alpha+\delta$ eutectoids	Dendritic with very few $\alpha+\delta$ eutectoids	Dendritic with a few $\alpha+\delta$ eutectoids
88	10	2	Dendritic with some $\alpha+\delta$ eutectoids	Dendritic with $\alpha+\delta$ eutectoids in the interdendritic regions	Dendritic with $\alpha+\delta$ eutectoids; long dendrites

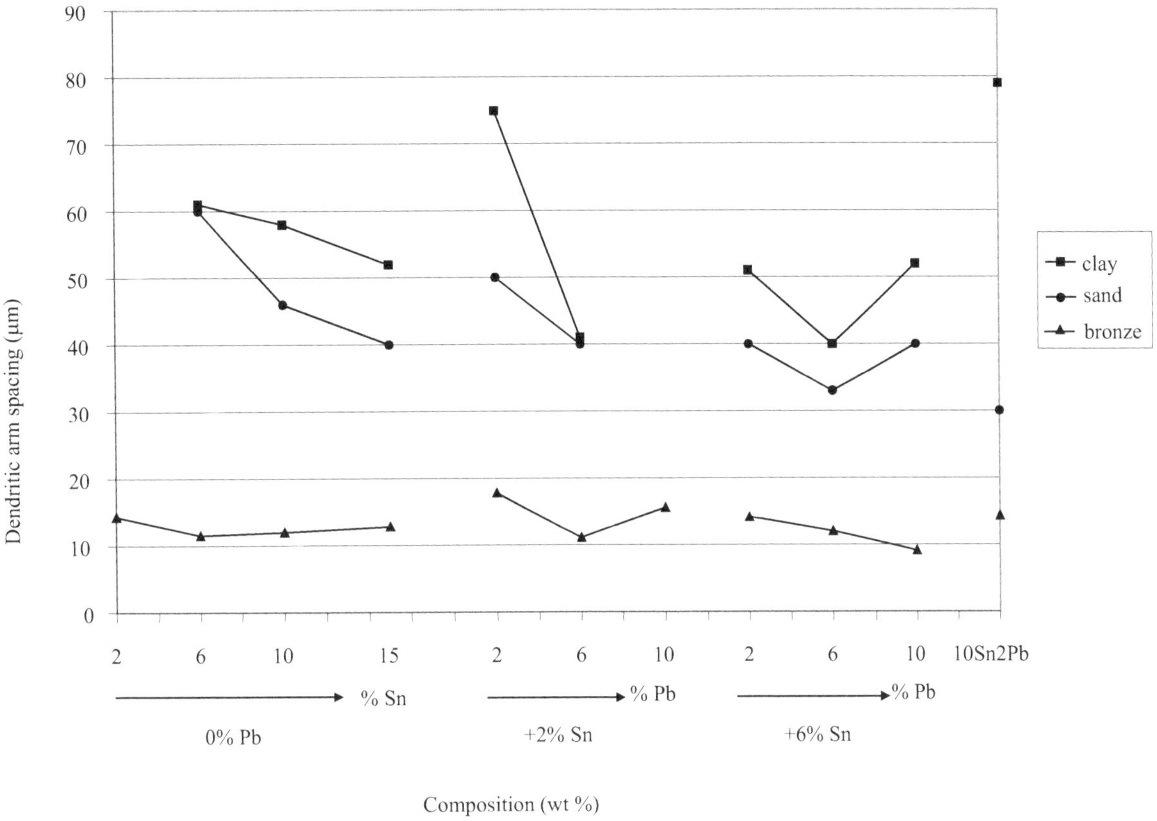

Figure 6.1 Dendritic arm spacing of bronzes cast in three different moulds, all air-cooled, showing the order: $DAS_{clay} > DAS_{sand} > DAS_{bronze}$

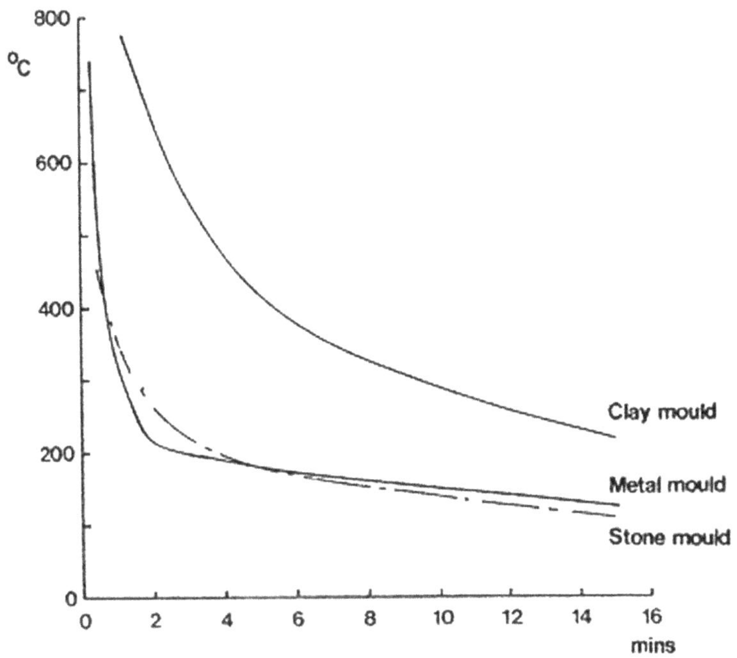

Figure 6.2 Cooling rate for a Cu + 10% Pb + 10% Sn bronze cast in metal, stone and clay moulds preheated to 100°C and air-cooled after casting (after Staniaszek and Northover 1983, 263)

Table 6.2 summarises the microstructures of water-quenched bronzes cast in sand and clay moulds. It can be seen that the amount of tin at which the needle β phase begins to appear is lower in bronzes cast in clay moulds (10%) than in bronzes cast in sand moulds (15%). This is also true of the point at which α+δ eutectoids begin to appear, which is lower in bronzes cast in clay moulds (2%) than in those cast in sand moulds (6%).

The dendritic arm spacing of the bronzes cast in sand and clay moulds and water-quenched is shown in figure 6.3. No consistent correlation between DAS values and the composition of bronzes cast in these moulds is indicated by the results.

Table 6.2 ***Microstructures of water-quenched axes cast in sand and clay moulds***

Composition (wt %)			**Moulding materials**	
Cu	**Sn**	**Pb**	**Sand**	**Clay**
98	2	—	Granular without α+δ eutectoids	Granular with α+δ eutectoids on grain boundaries
94	6	—	Dendritic with α+δ eutectoids	Dendritic with α+δ eutectoids in the interdendritic regions, no β
90	10	—	Dendritic with α+δ eutectoids; δ dark	Dendritic with β infill between dendrites
85	15	—	Dendritic with many α+δ eutectoids; δ on boundaries and needle β within grains	Dendritic with β infill between dendrites and within α grains
77	23	—	Granular, dark δ on boundaries and needle β within grains	β matrix with islands of α and δ
96	2	2	Both dendritic and granular without α+δ eutectoids	Dendritic with dark phase (probably δ) in the interdendritic regions
92	2	6	Both dendritic and granular, without α+δ eutectoids	Dendritic without α+δ eutectoids
88	2	10	Granular without α+δ eutectoids	Granular without α+δ eutectoids; Pb both on the grain boundaries and within the grains
92	6	2	Dendritic with α+δ eutectoids; dark δ	Dendritic, eutectoids with dark δ phase in the interdendritic regions
88	6	6	Dendritic with a few α+δ eutectoids	Dendritic, eutectoids with dark δ phase in the interdendritic regions
84	6	10	Dendritic without α+δ eutectoids	Dendritic, eutectoids with dark δ phase in the interdendritic regions
88	10	2	Dendritic with many α+δ eutectoids; dark δ	Dendritic, eutectoids with dark δ phase in the interdendritic regions; β possibly present within the eutectoids

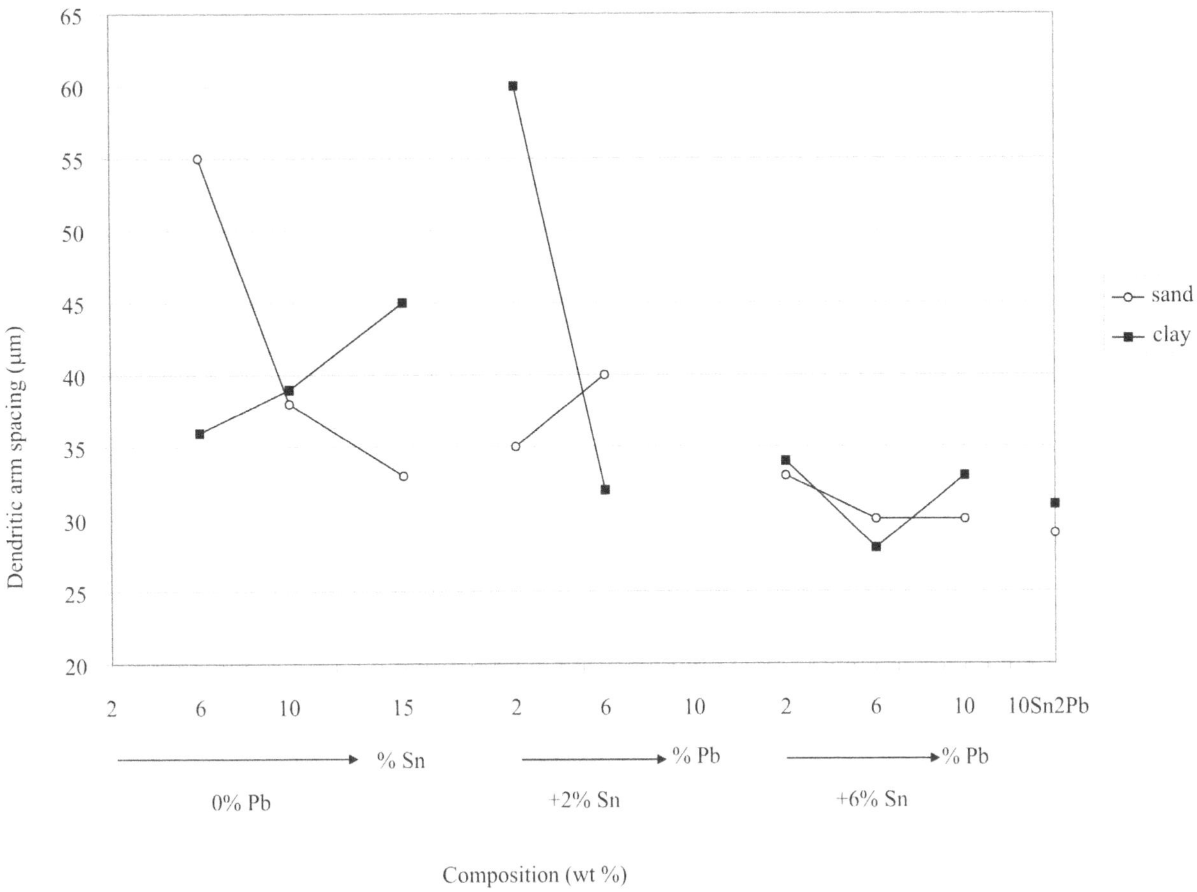

Figure 6.3 Dendritic arm spacing of water-quenched bronzes cast in sand and clay moulds

Table 6.3 summarises the microstructure of bronzes cast in preheated clay and bronze moulds. There is little difference in microstructure between casts made in moulds of different materials except for the casts with 2% tin and 6% lead, in which the cast made in a clay mould has a granular structure, while the cast made in a bronze mould has a dendritic structure.

The dendritic arm spacing of the casts made in the preheated moulds of clay and bronze is shown in figure 6.4, which indicates a considerably bigger dendritic arm spacing for the casts made in clay moulds than for those made in bronze moulds for each composition.

Table 6.3 ***Microstructures of axes cast in preheated clay and bronze moulds***

Composition (wt %)			**Moulding material**	
Cu	**Sn**	**Pb**	**Clay**	**Bronze**
98	2	—	Granular without α+δ eutectoids	Dendritic without α+δ eutectoids
94	6	—	Dendritic with very few α+δ eutectoids	Dendritic with α+δ eutectoids on the grain boundaries
90	10	—	Dendritic with α+δ eutectoids	Dendritic with many α+δ eutectoids
85	15	—	Dendritic with massive α+δ eutectoids	Dendritic with many α+δ eutectoids
77	23	—	Dendritic with massive α+δ eutectoids	Dendritic with massive α+δ eutectoids

Table 6.3 (contd.) ***Microstructures of axes cast in preheated clay and bronze moulds***

96	2	2	Dendritic without α+δ eutectoids	Dendritic without α+δ eutectoids
92	2	6	Granular without α+δ eutectoids; Pb on grain boundaries;	Dendritic without α+δ eutectoids; short dendrites
88	2	10	Granular without α+δ eutectoids; Pb both on the grain boundaries and within the grains	Dendritic without α+δ eutectoids; short dendrites
92	6	2	Dendritic with very few α+δ eutectoids	Dendritic without α+δ eutectoids
88	6	6	Dendritic with very few α+δ eutectoids	Dendritic with a few α+δ eutectoids; short dendrites
84	6	10	Dendritic with very few α+δ eutectoids	Dendritic with a few α+δ eutectoids
88	10	2	Dendritic with α+δ eutectoids in the interdendritic regions	Dendritic with many α+δ eutectoids; long dendrites

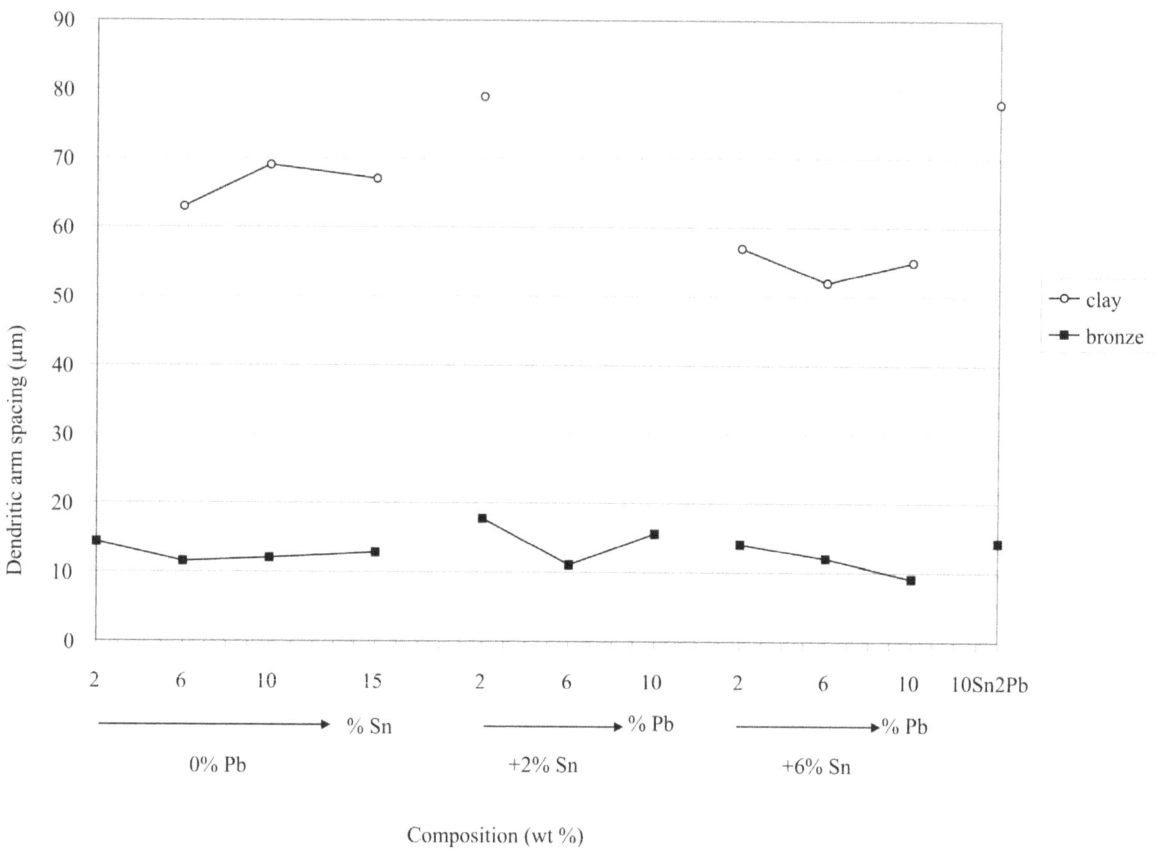

Figure 6.4 Dendritic arm spacing of bronzes cast in preheated clay and bronze moulds

In general these results support earlier observations that bronzes of various composition cast in clay moulds have a consistently larger dendritic arm spacing than those of the same composition cast in other moulds.

6.3 Microhardness

The microhardness values of as-cast bronzes cast in the three different moulding materials and air-cooled are presented in table 6.4 and illustrated in figure 6.5. It can be seen that, in agreement with well-established results, the hardness increases steeply with increasing tin content. The general variation in hardness values in the casts from different moulding materials is not very significant, although most casts produced in the bronze moulds are slightly harder. This can be explained by the larger number of $\alpha+\delta$ eutectoids in the bronze mould castings than in the other casts as the δ phase contains more tin and is hard and brittle.

The microhardness values of bronzes cast in sand and clay moulds and quenched in water are listed in table 6.5 and illustrated in figure 6.6. Most bronzes cast in clay moulds are slightly harder than those made in sand moulds.

The microhardness values of bronzes cast in preheated clay and bronze moulds and cooled in air are listed in table 6.6 and illustrated in figure 6.7. The results indicate that most bronzes cast in bronze moulds are slightly harder than those made in clay moulds

Hardness values quoted in publications by different scholars are notoriously difficult to compare. Tylecote (1987, 248) quotes a hardness of 75 Hv for a pure as-cast copper, while Scott (1991, 82) quotes 40-50 Hv for the same material, which agrees with a value of 50 Hv quoted by Chadwick (1939, 344) for the same material. Chadwick also gives a value of 120 Hv for a 15% tin-copper alloy. The hardness values presented here for the experimental castings agree with those quoted by Scott and Chadwick.

Table 6.4 ***Microhardness of air-cooled bronzes cast in three different moulding materials***

			Sand		Clay		Bronze	
Cu%	Sn%	Pb%	Hv	SD	Hv	SD	Hv	SD
98	2	0	62.9	1.0	60.4	5.2	57.9	7.4
94	6	0	77.7	6.8	77.6	7.1	82.0	4.1
90	10	0	99.9	4.4	88.9	3.5	112.2	12.7
85	15	0	126.8	21.5	124.8	10.7	159.4	7.0
77	23	0	219.4	20.3	257.8	33.7	238.6	14.3
96	2	2	58.5	2.0	54.6	4.7	70.0	9.2
92	2	6	54.3	10.2	57.1	5.9	59.6	5.4
88	2	10	54.5	3.6	59.5	3.5	55.9	0.4
92	6	2	74.5	8.4	74.4	10.5	82.6	1.5
88	6	6	69.6	1.7	70.7	10.0	85.6	4.9
84	6	10	68.4	5.7	75.2	7.3	79.3	4.2
88	10	2	91.8	6.0	96.7	14.7	108.4	13.4

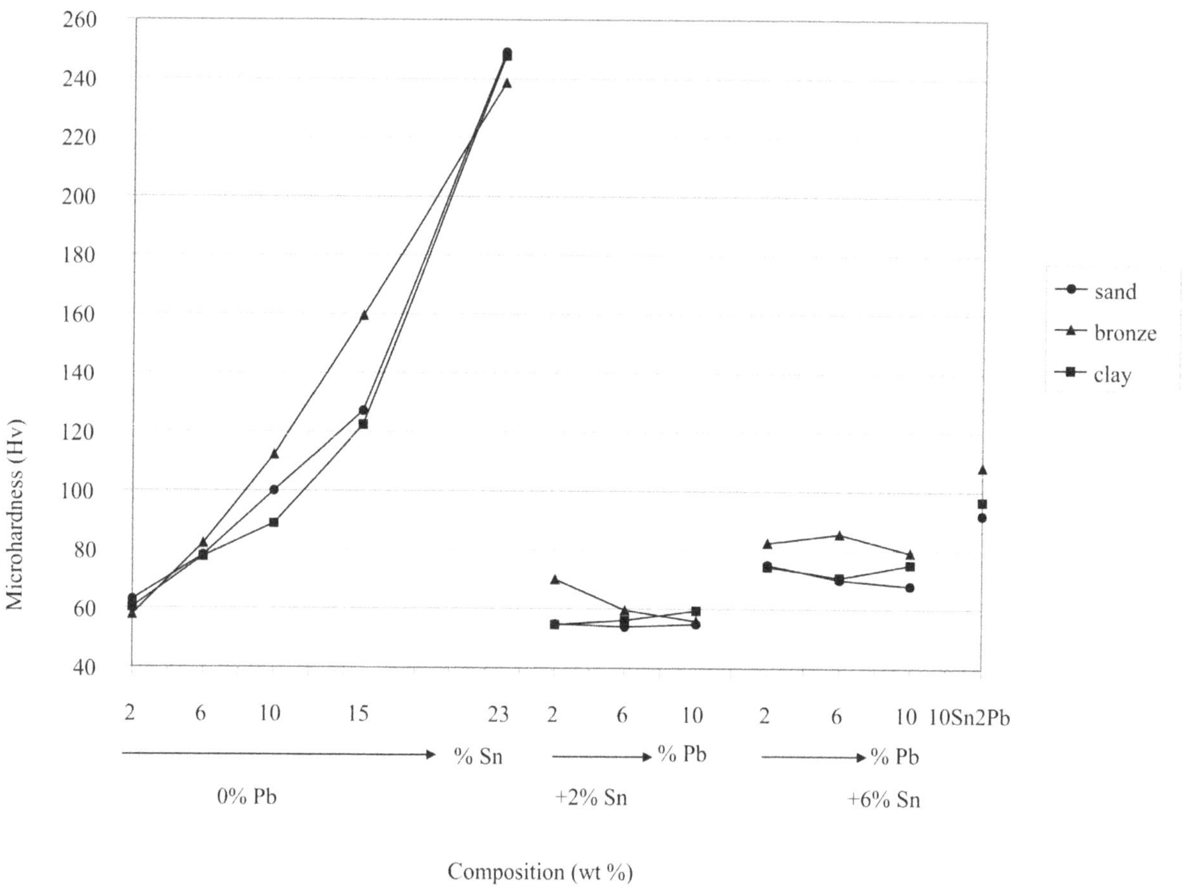

Figure 6.5 Vickers microhardness of air-cooled bronzes cast in three moulding materials

Table 6.5 ***Microhardness of water-quenched bronzes cast in sand and clay moulds***

Composition (wt %)			Sand		Clay	
Cu %	Sn %	Pb %	Hv	SD	Hv	SD
98	2	0	64.1	2.0	71.1	2.5
94	6	0	74.8	7.2	90.4	8.6
90	10	0	104.0	5.7	108.1	12.8
85	15	0	153.0	24.1	150.0	23.9
77	23	0	263.6	25.7	250.0	26.0
96	2	2	57.9	1.9	66.2	4.0
92	2	6	57.1	6.5	59.1	4.3
88	2	10	55.2	2.2	60.9	4.7
92	6	2	84.2	5.5	107.3	20.9
88	6	6	68.0	3.7	82.3	9.5
84	6	10	78.6	6.9	74.4	11.3
88	10	2	105.1	11.1	133.4	19.0

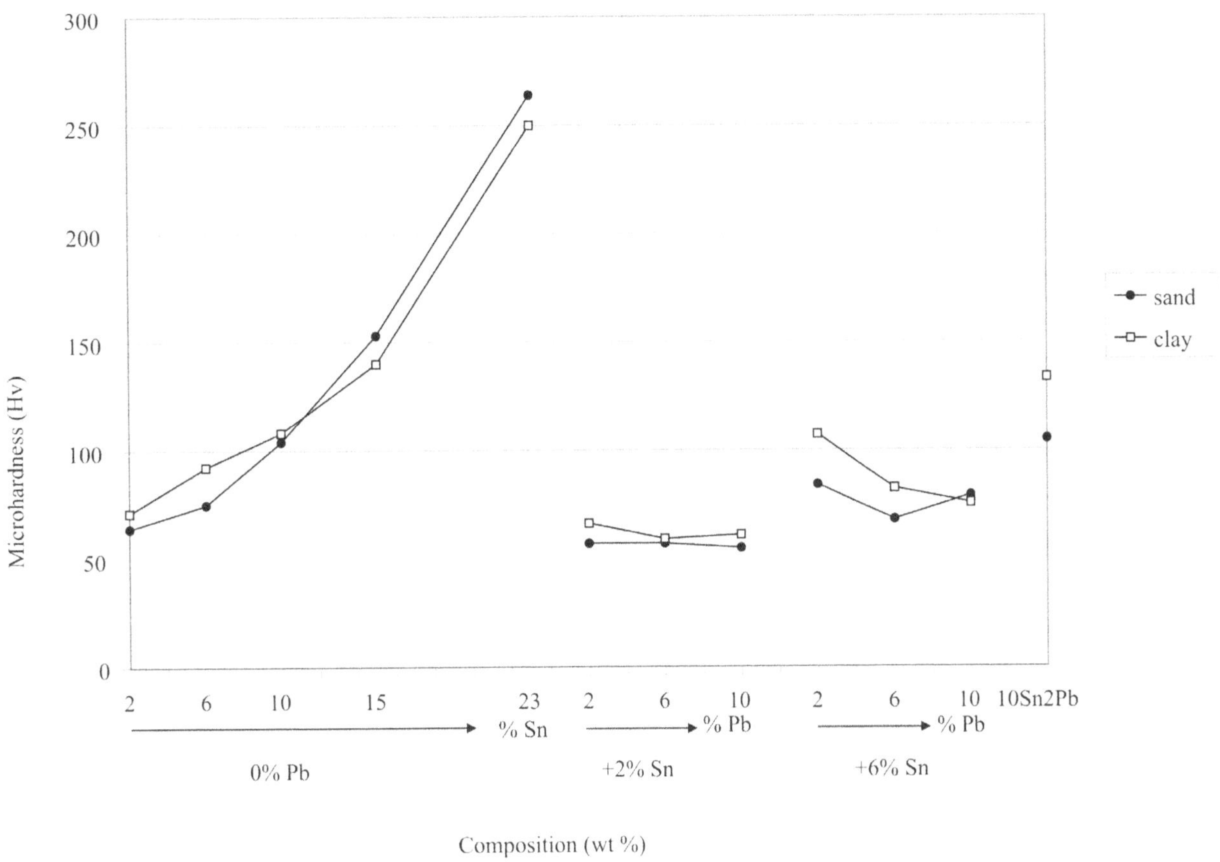

Figure 6.6 Vickers microhardness of water-quenched casts made in sand and clay moulds

Table 6.6 ***Microhardness of bronzes cast in preheated clay and bronze moulds and cooled in air***

Composition (wt %)			Clay		Bronze	
Cu %	**Sn %**	**Pb %**	**Hv**	**SD**	**Hv**	**SD**
98	2	0	59.6	2.7	63.1	5.0
94	6	0	75.4	8.7	78.8	9.2
90	10	0	99.4	10.9	117.6	10.1
85	15	0	109.2	11.0	138.0	6.7
77	23	0	225.4	44.4	247.6	20.1
96	2	2	60.7	1.8	58.8	1.9
92	2	6	52.1	4.6	58.0	3.9
88	2	10	51.9	10.5	59.7	6.5
92	6	2	81.7	5.9	76.3	2.0
88	6	6	74.6	1.9	77.6	2.7
84	6	10	72.1	7.9	73.7	4.1
88	10	2	101.2	6.6	96.3	6.2

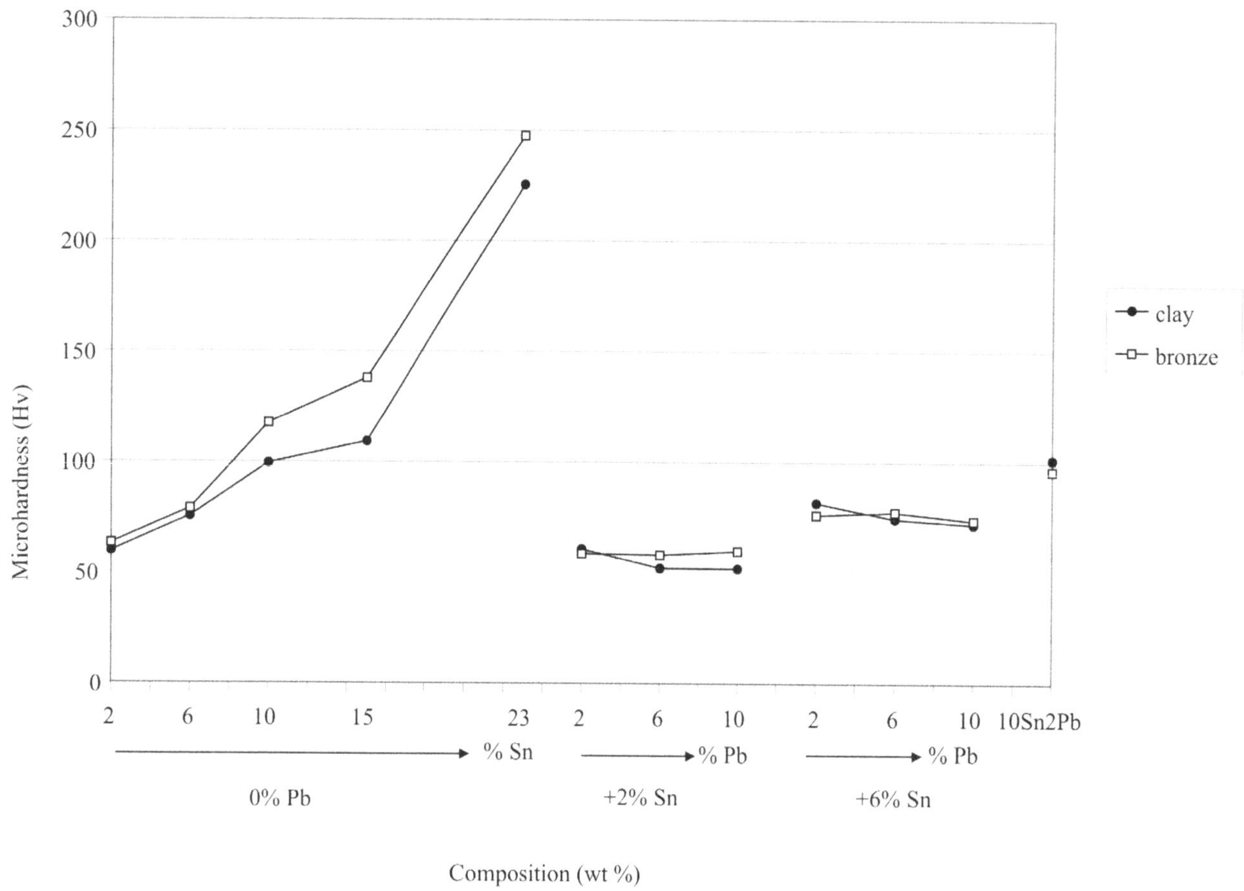

Figure 6.7 Vickers microhardness of bronzes cast in preheated clay and bronze moulds

6.4 Conclusions

The differences in microstructure of the bronzes cast in sand and clay moulds are not very significant. However, the microstructure of the bronzes cast in bronze moulds differs greatly, with a significantly smaller dendritic arm spacing than in those cast in sand and clay moulds.

Variations in microstructure have been observed for alloys of different composition. Most bronzes containing 2% tin and cast in sand and clay moulds have a granular structure, while those cast in bronze moulds have a dendritic structure, at least at the surface. There are no $\alpha+\delta$ eutectoids in any of these bronzes with 2% tin, irrespective of either mould material or lead content. All casts with 6% tin have a dendritic structure, irrespective of mould material or lead content. $\alpha+\delta$ eutectoids begin to appear in bronzes with a tin content of 6% or higher, irrespective of mould material. However, very few or no $\alpha+\delta$ eutectoids are present in leaded bronze with 6% tin. This indicates that the addition of lead does have an effect on the microstructure although, in theory, lead is immiscible with copper and thus should not affect the microstructure. However, as lead lowers the melting-point of an alloy, possibly changing the casting temperature, the microstructure is consequently affected. This is most noticeable in 2% tin bronzes with lead, in which bronze with a higher lead content (10%) has a granular structure and that with lower lead has a mixed granular and dendritic microstructure. All bronzes with a tin content of 10% and higher have a dendritic structure with $\alpha+\delta$ eutectoids. The higher the tin content, the more $\alpha+\delta$ eutectoids there are.

The microhardness increases greatly, as is well known, with increasing tin content. There is little difference in hardness between casts made in clay and sand moulds. Most casts produced in bronze moulds are slightly, but consistently, harder than those cast in either clay or sand moulds.

7 Comparison of the malleability of bronzes cast in three different moulding materials with different cooling rates

7.1 Introduction

To investigate the workability of the bronzes cast in this series, the following experiments were carried out.

1. Two bronzes with 6% tin cast in sand moulds, one cooled in air and the other quenched in water, were chosen to study the effect of cooling rates on workability.
2. Bronzes containing 6% tin cast in three different moulding materials, all air-cooled, were chosen for the study of the effect of moulding material on workability.
3. Bronzes of various compositions cast in sand moulds and cooled in air were chosen for studying the effect of alloying composition on workability.

Reduction in thickness of the metals was achieved by cold-rolling. The reduction refers to the total (accumulative) reduction rather than the reduction after the last step. After each reduction a sample was taken, mounted in resin, polished and etched in preparation for taking photomicrographs and hardness measurements. A further cold-rolling was then carried out on the remainder and the sample preparation was repeated. Microhardness was measured as described in chapter 2.3. Five readings were taken for most samples except where the sample was too small. Table 7.1 gives the basic data for this chapter: it includes the composition of the samples, the cooling method employed after casting, whether work was carried out on the samples (post-casting treatment), the resulting reduction and the hardness measurement for all samples discussed in this chapter. Some of the samples were annealed after cold-working to enable further cold-working to be carried out. These samples are denoted 'a-cw' in table 7.1. Details of annealing, such as temperature and duration, will be discussed in section 7.5.

Table 7.1 ***Composition, cooling method, reduction and microhardness values (Vickers scale Hv) of samples discussed in chapter 7***

Sample no.	Sn %	Pb %	Cooling method	Post-casting treatment [1]	Reduction (%)	Hv1	Hv2	Hv3	Hv4	Hv5	Hv mean	SD
Sand mould												
41	2	0	water	as	0	63.6	63.6	61.3	66.0	66.0	64.1	2.0
42	2	0	air	as	0	63.6	63.6	61.3	62.4	63.6	62.9	1.0
				cw	10	95.8	82.2	82.2	98.0	82.2	88.0	8.1
				cw	30	100.0	136.0	122.0	139.0	139.0	127.0	16.8
				a-cw	45	119.0	122.0	113.0	136.0	139.0	126.0	11.2
				a-cw	60	100.0	136.0	125.0	139.0	119.0	124.0	15.6
35	6	0	water	as	0	71.6	64.8	82.9	74.2	80.5	74.8	7.2
35I				cw	10	128.0	122.0	122.0	128.0	128.0	125.6	3.3
35II				cw	30	151.0	136.0	122.0	156.0	156.0	144.2	14.9
35III				cw	45	181.0	199.0	151.0	187.0	181.0	179.8	17.7
35IV				cw	60	213.0	206.0	165.0	170.0	176.0	186.0	21.9
35V				cw	80	221.0	221.0	199.0	187.0	199.0	205.4	15.1
35VI				cw	90	236.0	187.0	187.0	245.0	221.0	215.2	27.1
35VII				cw	95	228.0		228.0		228.0	228.0	0.0
36	6	0	air	as	0	76.0	67.3	78.8	85.8	80.5	77.7	6.8
36I				cw	10	132.0	119.0	91.6	132.0	108.0	116.5	17.2
36II				cw	30	160.0	122.0	139.0	122.0	160.0	140.6	19.0
36III				cw	45	193.0	181.0	199.0	181.0	181.0	187.0	8.5
36IV				cw	60	187.0	176.0	199.0		221.0	195.8	19.3
36V				cw	80	221.0	199.0	193.0	199.0	228.0	208.0	15.5
36VI				cw	90	236.0		187.0		199.0	207.3	25.5
43	10	0	water	as	0	108.0	108.0	95.8	108.0	100.0	104.0	5.7
44	10	0	air	as	0	100.0	103.0	93.6	98.0	105.0	99.9	4.4

Table 7.1 (contd.) ***Composition, cooling method, reduction and microhardness values (Vickers scale Hv) of samples discussed in chapter 7***

				cw	10	128.0	110.0	125.0	139.0	105.0	121.0	13.8
				cw	30	206.0	156.0	199.0	176.0	165.0	180.0	21.5
				a-cw	45	213.0	187.0	187.0	165.0	193.0	189.0	17.1
				a-cw	60	199.0	187.0	181.0	170.0	181.0	184.0	10.6
37	15	0	water	as	0	153.0	153.0	116.0	183.0	160.0	153.0	24.1
38	15	0	air	as	0	111.0	118.0	115.0	126.0	164.0	126.8	21.5
				cw	10	193.0	193.0	165.0	187.0	156.0	179.0	17.2
				cw	30	228.0	199.0	206.0	206.0	228.0	213.0	13.6
				a-cw	45	193.0	193.0	176.0	165.0	193.0	184.0	12.9
39	23	0	water	as	0	225.0	262.0	268.0	266.0	297.0	263.6	25.7
40	23	0	air	as	0	224.0	264.0	236.0	249.0	274.0	249.4	20.3
45	2	2	water	as	0	57.1	57.1	57.1	61.3	57.1	57.9	1.9
46	2	2	air	as	0	59.1	59.1	60.2	59.1	55.1	58.5	2.0
				cw	10	93.6	75.5	72.7	67.3	91.6	80.0	11.8
				cw	30	125.0	98.0	100.0	93.6	122.0	108.0	14.6
				a-cw	45	119.0	128.0	116.0	110.0	128.0	120.0	7.8
				a-cw	60	119.0	139.0	136.0	108.0	125.0	125.0	12.7
47	2	6	water	as	0	46.0	59.1	63.3	59.1	58.1	57.1	6.5
48	2	6	air	as	0	69.9	55.1	46.7	43.9	56.1	54.3	10.2
				cw	10	74.2	74.2	66.0	84.0	93.6	78.0	10.6
				cw	30	100.0	98.0	89.6	105.0	119.0	102.0	10.9
				a-cw	45	139.0	103.0	91.6	105.0	108.0	109.0	17.7
				a-cw	60	125.0	128.0	119.0	122.0	132.0	125.0	5.1
49	2	10	water	as	0	56.1	52.4	57.1	57.1	53.3	55.2	2.2
50	2	10	air	as	0	58.1	53.3	58.1	53.3	49.8	54.5	3.6
				cw	10	64.8	53.3	45.3	62.4	77.2	61.0	12.1
				cw	30	85.8	103.0	68.6	97.6	116.0	94.0	18.0
				a-cw	45	108.0	116.0	91.6	110.0	103.0	106.0	9.2
				a-cw	60	84.0	95.8	125.0	122.0	98.0	105.0	17.8
51	6	2	water	as	0	80.5	89.6	89.6	77.2	84.0	84.2	5.5
52	6	2	air	as	0	74.2	84.0	74.2	61.3	78.8	74.5	8.4
				cw	10	89.6	71.3	89.6	93.6	110.0	91.0	13.8
				cw	30	125.0	119.0	128.0	108.0	132.0	122.0	9.3
				a-cw	45	113.0	125.0	119.0	160.0	160.0	135.0	22.9
				a-cw	60	160.0	160.0	143.0	151.0	193.0	161.0	19.0
53	6	6	water	as	0	63.6	72.7	69.9	64.8	68.8	68.0	3.7
54	6	6	air	as	0	68.8	72.7	68.8	68.8	68.8	69.6	1.7
				cw	10	78.8	78.8	68.6	75.5	85.8	78.0	6.2
				cw	30	128.0	95.8	122.0	95.8	119.0	112.0	15.2
				a-cw	45	151.0	116.0	132.0	151.0	125.0	135.0	15.7
				a-cw	60	165.0	143.0	136.0	122.0	110.0	135.0	21.0
55	6	10	water	as	0	85.8	82.2	69.9	82.2	72.7	78.6	6.9
56	6	10	air	as	0	68.8	74.2	71.3	68.8	59.1	68.4	5.7
				cw	10	110.0	77.2	84.0	84.0	80.5	87.0	13.1
				cw	30	116.0	147.0	132.0	119.0	119.0	127.0	13.0
				a-cw	45	119.0	100.0	103.0	82.2	122.0	105.0	16.1
				a-cw	60	160.0	119.0	139.0	139.0	136.0	139.0	14.6
57	10	2	water	as	0	122.0	93.6	110.0	100.0	100.0	105.1	11.1
58	10	2	air	as	0	100.0	93.6	85.8	93.6	85.8	91.8	6.0
				cw	10	110.0	128.0	113.0	139.0	151.0	128.0	17.3
				cw	30	193.0	160.0	170.0	128.0	147.0	160.0	24.4
				a-cw	45	181.0	181.0	156.0	181.0	176.0	175.0	10.8
				a-cw	60	147.0	170.0	156.0	176.0	122.0	154.0	21.3

Table 7.1 (contd.) ***Composition, cooling method, reduction and microhardness values (Vickers scale Hv) of samples discussed in chapter 7***

Clay mould

111	2	0	preheated	as	0	61.3	61.3	61.3	59.1	55.1	59.6	2.7
112	2	0	air	as	0	62.4	56.1	68.6	57.5	57.5	60.4	5.2
113	2	0	water	as	0	71.3	74.2	71.3	71.3	67.3	71.1	2.5
114	6	0	preheated	as	0	77.2	67.3	66.7	78.1	87.6	75.4	8.7
115	6	0	air	as	0	80.5	74.2	68.6	77.2	87.6	77.6	7.1
115I				cw	10	100.0	72.7	87.6	85.8	139.0	97.0	25.4
115II				cw	30	181.0	147.0	119.0	136.0	128.0	142.2	24.0
115III				cw	45	181.0	136.0	147.0	147.0	143.0	150.8	17.5
115IV				cw	60	181.0	193.0	206.0	206.0	208.0	198.8	11.6
115V				cw	80	193.0	199.0	170.0	236.0	181.0	195.8	25.1
116	6	0	water	as	0	84.0	84.0	105.0	89.6	89.6	90.4	8.6
117	10	0	preheated	as	0	116.0	97.5	98.0	85.5	100.0	99.4	10.9
118	10	0	air	as	0	84.0	87.6	89.6	89.6	93.6	88.9	3.5
119	10	0	water	as	0	122.0	116.0	98.0	113.0	91.6	108.1	12.8
120	15	0	preheated	as	0	121.0	110.0	98.0	98.0	119.0	109.2	11.0
121	15	0	air	as	0	135.0	129.0	108.0	121.0	131.0	124.8	10.7
122	15	0	water	as	0	174.0	113.0	168.0	149.0	146.0	150.0	23.9
123	23	0	preheated	as	0	176.0	213.0	213.0	297.0	228.0	225.4	44.4
124	23	0	air	as	0	279.0	233.0	264.0	215.0	298.0	257.8	33.7
125	23	0	water	as	0	294.0	251.0	227.0	238.0	240.0	250.0	26.0
126	2	2	preheated	as	0	62.7	61.5	60.0	61.3	57.9	60.7	1.8
127	2	2	air	as	0	55.5	50.7	57.1	49.2	60.6	54.6	4.7
128	2	2	water	as	0	71.3	63.6	68.6	61.3	66.0	66.2	4.0
129	2	6	preheated	as	0	55.1	55.1	47.5	56.1	46.7	52.1	4.6
130	2	6	air	as	0	66.8	54.2	53.8	58.3	52.4	57.1	5.9
131	2	6	water	as	0	56.1	60.2	53.6	64.3	61.3	59.1	4.3
132	2	10	preheated	as	0	48.2	62.4	40.1	63.6	45.3	51.9	10.5
133	2	10	air	as	0	54.4	62.4	62.4	57.3	61.1	59.5	3.5
134	2	10	water	as	0	66.0	56.1	61.3	56.1	64.8	60.9	4.7
135	6	2	preheated	as	0	84.7	72.7	78.8	85.8	86.5	81.7	5.9
136	6	2	air	as	0	80.5	74.2	60.2	69.4	87.6	74.4	10.5
137	6	2	water	as	0	143.0	107.0	98.8	89.6	98.0	107.3	20.9
138	6	6	preheated	as	0	74.2	71.9	77.2	75.4	74.2	74.6	1.9
139	6	6	air	as	0	74.2	55.1	67.3	76.6	80.5	70.7	10.0
140	6	6	water	as	0	96.7	86.9	73.6	78.8	75.5	82.3	9.5
141	6	10	preheated	as	0	71.3	64.8	70.5	85.5	68.6	72.1	7.9
142	6	10	air	as	0	84.0	69.9	69.9	82.2	69.9	75.2	7.3
143	6	10	water	as	0	71.3	58.3	87.3	72.1	82.9	74.4	11.3
144	10	2	preheated	as	0	105.0	93.6	108.0	94.6	105.0	101.2	6.6
145	10	2	air	as	0	98.0	82.2	91.6	90.8	121.0	96.7	14.7
146	10	2	water	as	0	122.0	108.0	136.0	157.0	144.0	133.4	19.0

Bronze mould

203	2	0	preheated	as	0	71.3	58.1	61.3	61.3	63.6	63.1	5.0
204	2	0	air	as	0	62.4	45.3	63.6	60.2	58.1	57.9	7.4
205	6	0	preheated	as	0	80.5	69.9	69.9	82.2	91.6	78.8	9.2
206	6	0	air	as	0	82.2	87.6	78.8	84.0	77.2	82.0	4.1
206I				cw	10	108.0	100.0	95.8	116.0	119.0	107.8	10.0
206II				cw	30	151.0	151.0	176.0	160.0	165.0	160.6	10.5
206III				cw	45	213.0	181.0	156.0	139.0	193.0	176.4	29.4
206IV				cw	60	213.0	187.0	151.0	187.0	187.0	185.0	22.1
206V				cw	80	221.0	199.0		187.0	181.0	197.0	17.7
207	10	0	preheated	as	0	100.0	122.0	119.0	122.0	125.0	117.6	10.1

Table 7.1 (contd.) ***Composition, cooling method, reduction and microhardness values (Vickers scale Hv) of samples discussed in chapter 7***

208	10	0	air	as	0	110.0	103.0	100.0	116.0	132.0	112.2	12.7
209	15	0	preheated	as	0	147.0	132.0	132.0	143.0	136.0	138.0	6.7
210	15	0	air	as	0	156.0	160.0	160.0	151.0	170.0	159.4	7.0
211	23	0	preheated	as	0	264.0	236.0	236.0	228.0	274.0	247.6	20.1
212	23	0	air	as	0	245.0	213.0	245.0	245.0	245.0	238.6	14.3
213	2	2	preheated	as	0	61.3	57.1	58.1	60.2	57.1	58.8	1.9
214	2	2	air	as	0	85.8	63.6	63.6	68.6	68.6	70.0	9.2
215	2	6	preheated	as	0	55.1	58.1	64.8	56.1	56.1	58.0	3.9
216	2	6	air	as	0	59.1	52.4	58.1	67.3	61.3	59.6	5.4
217	2	10	preheated	as	0	68.6	64.8	55.1	55.1	55.1	59.7	6.5
218	2	10	air	as	0	55.1	56.1	56.1	56.1	56.1	55.9	0.4
219	6	2	preheated	as	0	77.2	72.7	77.2	77.2	77.2	76.3	2.0
220	6	2	air	as	0	84.0	82.2	82.2	80.5	84.0	82.6	1.5
221	6	6	preheated	as	0	77.2	77.2	75.7	75.7	82.2	77.6	2.7
222	6	6	air	as	0	87.6	91.6	80.5	80.5	87.6	85.6	4.9
223	6	10	preheated	as	0	74.2	67.3	77.2	77.2	72.7	73.7	4.1
224	6	10	air	as	0	78.8	75.7	85.8	80.5	75.7	79.3	4.2
225	10	2	preheated	as	0	95.8	85.8	100.0	100.0	100.0	96.3	6.2
226	10	2	air	as	0	100.0	100.0	105.0	105.0	132.0	108.4	13.4

Note: [1] as – as-cast; cw – cold-worked; a – annealed

7.2 The effect of cooling rate

Cold-rolling was conducted on two 6% tin bronzes cast in sand moulds: one was air-cooled (no. 36) and the other was quenched in water (no. 35). Cracking occurred in the air-cooled one at a reduction of 90%, but was not apparent in the quenched bronze until it reached a reduction of 95% (table 7.1). The experiments therefore showed that quenched bronze was more malleable than air-cooled bronze.

The microhardness of the cold-worked 6% tin bronzes was measured at each stage of reduction, i.e. at 10, 30, 45, 60, 80, 90, and 95% for the water-quenched sample. The results are listed in table 7.1. Figure 7.1, in which the results are plotted, shows little difference in hardness between the bronzes cooled at different rates. In other words, the cooling method did not result in significantly different hardness values until the bronzes reached 90% reduction.

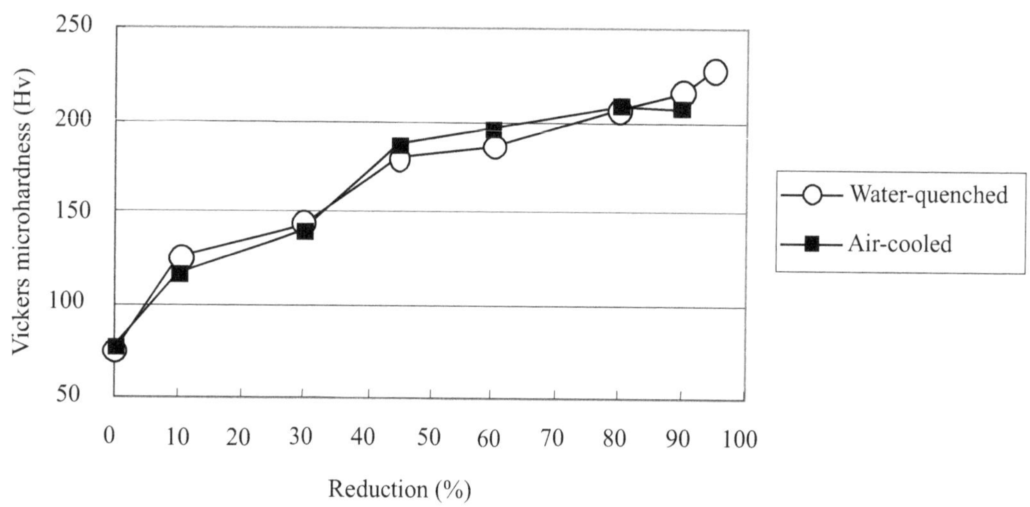

Figure 7.1 Microhardness of the cold-worked 6% Sn bronzes (cast in sand) at different reduction rates

7.3 The effect of moulding material

Reduction in thickness was conducted on 6% tin bronzes, cast in the three different moulding materials and air-cooled, by cold-rolling without any annealing until the bronzes cracked. The bronze cast in the sand mould, as observed above, did not crack until reduction reached 90%, while the bronzes cast in clay and bronze moulds cracked at a reduction of 80%. The microhardness values at each reduction are listed in table 7.1. They are plotted in figure 7.2, which seems to indicate a tendency for bronze cast in a clay mould to have lower hardness values than bronzes cast in the other moulding materials. However, in view of the large standard deviations of the hardness measurements, there is no clear or significant correlation between hardness and moulding materials.

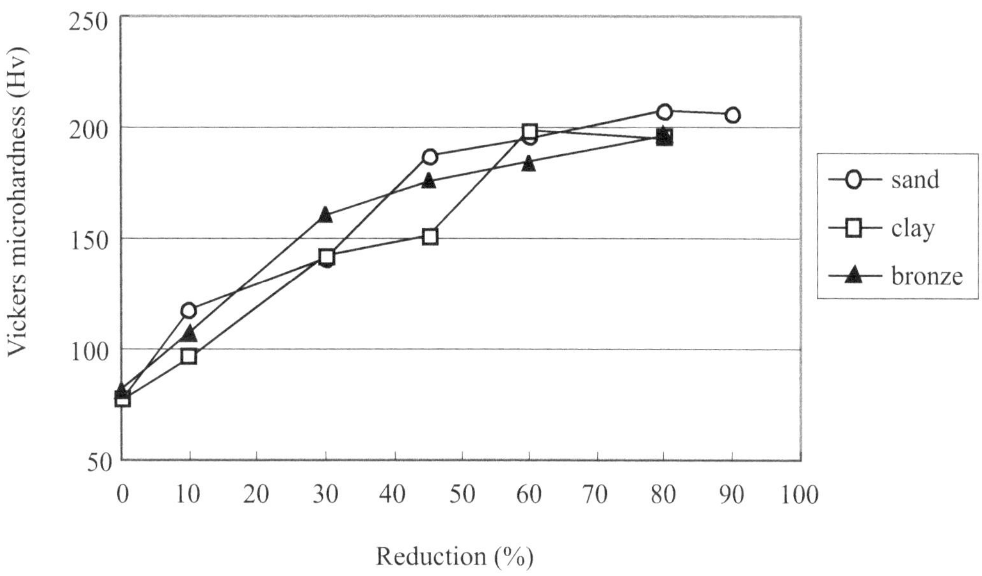

Figure 7.2 Microhardness of the cold-worked 6% Sn bronzes (cooled in air) cast in different moulding materials

7.4 The effect of alloying composition

Cold-working (i.e. cold-rolling) of air-cooled bronzes cast in sand moulds was carried out on all samples except the one with 23% tin, as a bronze with 17% or more tin cannot be cold-worked (Scott 1991, 26). Two reductions, to 10 and 30%, were done on each sample before annealing.

After the 30% reduction the samples were annealed and further reduction to a total of 45% was carried out. After this, the samples were annealed again and further reduced to a total of 60%. The procedure of reduction by cold-rolling and annealing is presented diagrammatically in figure 7.3, in which R refers to the total reduction. The annealing process will be discussed in the next section.

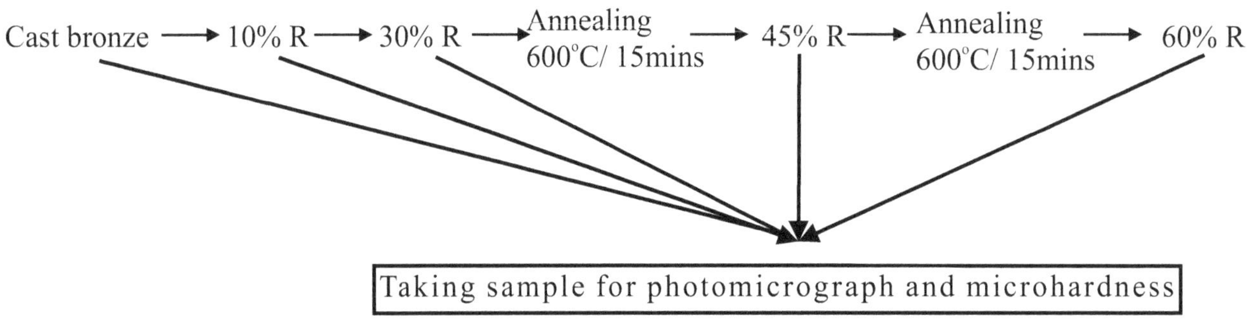

Figure 7.3 Diagrammatic representation of the reduction and annealing regime for air-cooled bronzes cast in sand moulds

The microhardness values of the samples at each reduction are listed in table 7.1. They are plotted against the alloy compositions in figure 7.4. Only reductions before annealing, i.e. the 10 and 30% reductions, have been taken into account here for investigating the effects of alloying composition on the workability of the bronzes, because the 45% and 60% reduction involved two processes, annealing and cold-working, which have opposite effects on hardness.

Figure 7.4 can be divided into three groups: bronzes with varying amounts of tin but no lead, and leaded bronzes with 2% and 6% tin, each group containing varying amounts of lead. In the tin bronze group, the hardness increased with the tin content at each reduction, with a trend parallel to that of the as-cast samples (0% reduction). The sample containing 10% tin, which had undergone 10% reduction, appears to be less hard than might be expected, but, as mentioned earlier, the hardness values do have large standard deviations and this experiment should be repeated before general conclusions can be drawn from it (table 7.1). The leaded bronzes containing 2% tin and 2, 6 and 10% lead show a decrease of hardness as the amount of lead increases, at both reductions. This trend is followed by the group of leaded bronzes containing 6% tin and 2 and 6% lead. The 10% lead, 6% Sn sample in this group has been left out of figure 7.4, as it was shown by ICP analysis to contain only 6.18% lead (see table 8.1, batch no. 56).

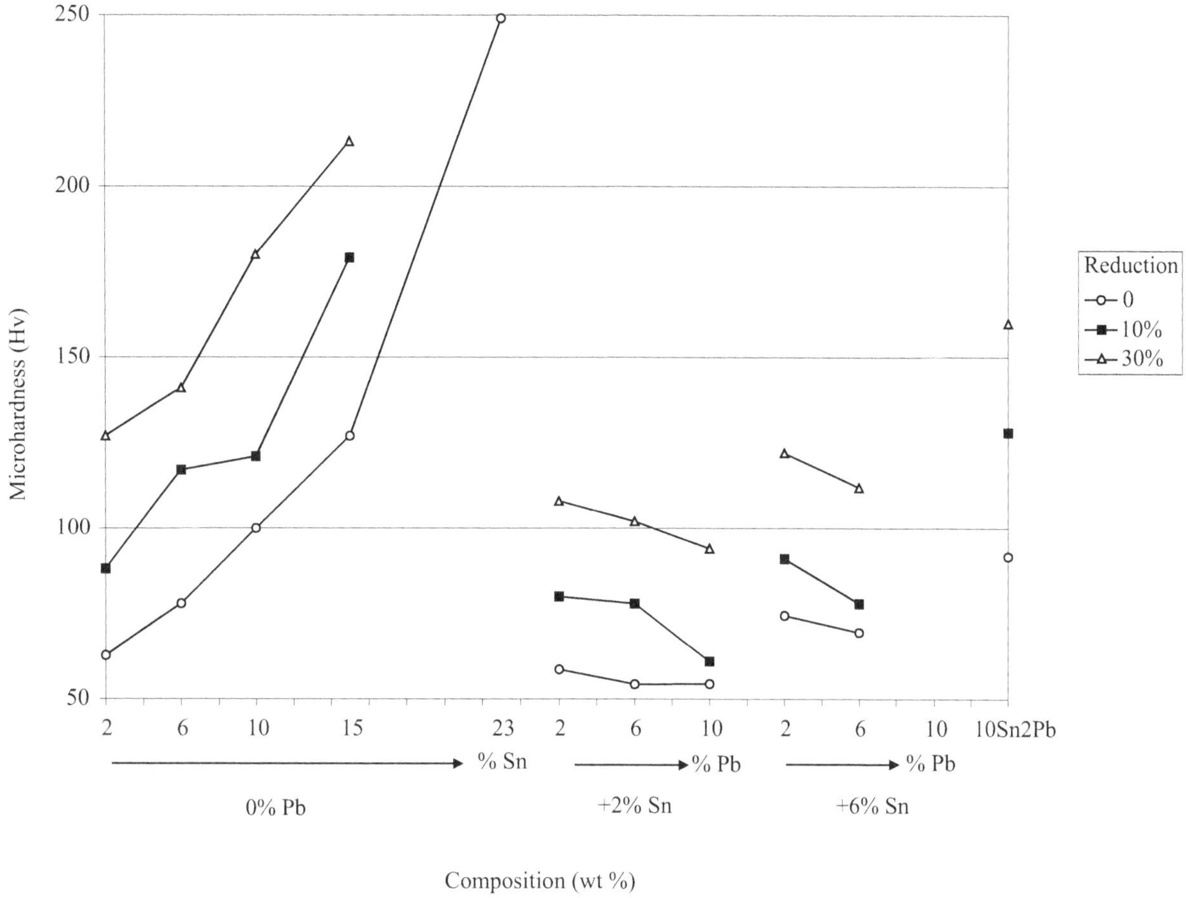

Figure 7.4 Vickers microhardness of cold-worked bronzes containing various contents of Sn and Pb cast in sand moulds and cooled in air

The bronze with 15% tin (no. 38) cracked at 30% reduction with a big crack running across the sample. The bronzes with 6% tin + 6% lead (no. 54), 6% tin + 10% lead (no. 56) and 10% tin + 2% lead (no. 58) had small cracks on the surface after the 30% reduction.

More cracks appeared in sample nos. 38, 54 and 56 at a reduction of 45%, whereas no cracks were visible in nos. 42 (2% Sn), 44 (10% Sn), 46, 48, 50 (2% Sn + variable Pb) and 52 (6% Sn +2% Pb) at 45% reduction. This confirms that the addition of lead does affect the malleability of bronzes. Table 7.2 summarises the visual effects of the cold-working and annealing cycles discussed in this chapter and the microstructure of the samples, which will be discussed in section 7.6, for each reduction.

Table 7.2 ***Summary of the cold-working/annealing cycle on air-cooled bronzes cast in sand moulds***

Sample no.	Sn %	Pb %	30%R	Annealed at 600°C/15 mins	45%R	Annealed at 650°C/15 mins	60%R
42	2	0		Original (granular) structure remaining	No cracks	Equi-axed	Micro-cracks near the surface
44	10	0		Dendritic structure disappeared but many α+δ eutectoids remaining	No cracks	Equi-axed with some remnant δ phase	No cracks
38	15	0	Cracks	Many α+δ eutectoids remaining	More cracks	Equi-axed with some remnant δ phase	Not conducted
46	2	2		Dendritic structure partially remaining	Micro-cracks starting to appear	Equi-axed with tin oxide and sulphide inclusions	Cracks
48	2	6		Dendritic structure partially remaining	No cracks	Equi-axed	Cracks
50	2	10		Dendritic structure remaining	No cracks	Equi-axed	Cracks
52	6	2		Both dendritic structure and α+δ eutectoids remaining	No cracks	Equi-axed	Cracks
54	6	6	Cracks visible on the surface but not in the polished section	Dendritic structure partially remaining	No cracks visible in the polished section	Equi-axed	Cracks
56	6	10 [1]	Cracks visible on the surface	Both dendritic structure and α+δ eutectoids remaining	More cracks on the surface	Equi-axed	Cracks
58	10	2	Cracks visible on the surface	Dendritic structure and α+δ eutectoids partially remaining	More cracks on the surface	Equi-axed	Cracks

Note: [1] Sample was shown to have only 6.18% Pb on ICP-OES analysis result.

7.5 Annealing

Annealing is the process by which the distorted cold-worked structure is changed by heating to a strain-free structure by recrystallisation. The recrystallisation process, which is dependent on the composition of the metal, is far more sensitive to changes in annealing temperature than to variations in time at constant temperature. The greater the amount of prior deformation, the lower the temperature at which recrystallisation starts (Avner 1974, 133). To test annealing temperatures of bronzes cast in this series, samples of the cold-worked (air-cooled) 6% tin bronze with 40% reduction were heated to 550, 600 and 650°C, respectively, for 15 minutes. The microstructure of all these annealed samples was fully recrystallised with annealing twins, as is illustrated by the 40% cold-worked, 6% tin bronze annealed at 550°C (fig. 7.5a) and the same bronze annealed at 650°C (fig. 7.5b).

The air-cooled 10% tin bronze has a smaller grain size than the air-cooled 6% tin bronze (see table 3.2). It is supposed to have a lower recrystallisation temperature because of its finer initial grain size (Avner 1974, 133). However, a cold-worked sample of an air-cooled 10% tin bronze which had undergone 30% reduction showed incomplete recrystallisation after having been annealed at 600°C for 15 minutes and α+δ eutectoids could still be seen in the microstructure (fig 7.6a). The structures of other samples at 30% reduction, which were annealed at 600°C for 15 minutes, such as 15% tin, 6% tin with 2, 6 and 10% lead, and 10% tin with 2% lead, were also incompletely recrystallised.

To make further observations, each cold-worked sample at 45% reduction was again annealed at 650°C for 15 minutes. The microstructures of the annealed samples showed that recrystallisation was almost complete, i.e. they all had equi-axed structures, although some δ phase remained in the high-tin bronzes (e.g. 10, 15% Sn) (fig. 7.6b). The remnant δ phase in the final annealed sample was much smaller than in the earlier one (fig. 7.6a). The results show that an increase in the amount of deformation prior to annealing does effectively decrease the recrystallisation temperature for a given annealing time. More tests on annealing temperature and time need to be carried out to obtain a complete picture of the complex behaviour of bronzes of archaeologically relevant composition, which may have been subjected to varying degrees of cold-working and annealing.

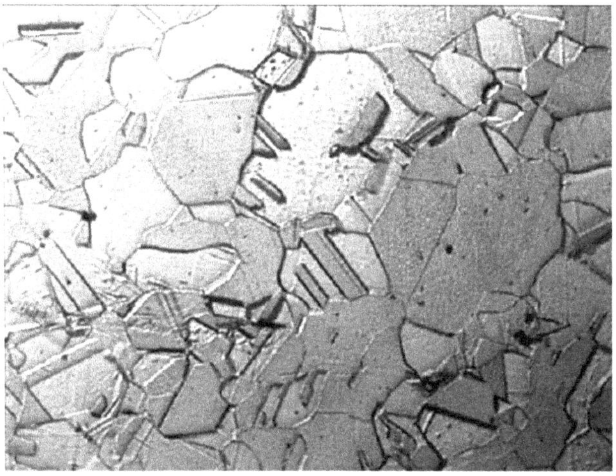

a. Annealed at 550°C for 15 minutes (image width 0.13 mm)

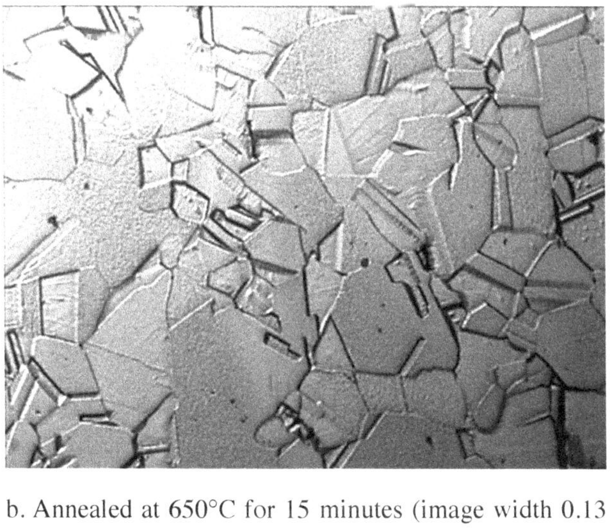

b. Annealed at 650°C for 15 minutes (image width 0.13 mm)

Figure 7.5 Photomicrographs of 40% cold-worked and annealed 6% Sn bronze (cast in sand and air-cooled), showing a fully recrystallised structure and annealing twins

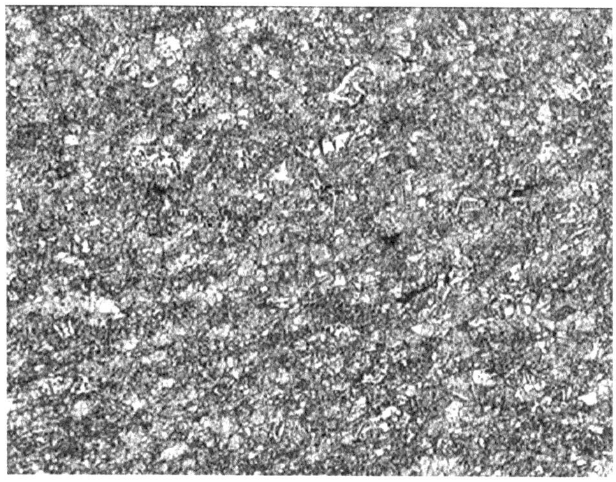

a. Annealed at 600°/15 minutes showing incomplete recrystallisation with considerable number of α+δ eutectoids (image width 0.13 mm)

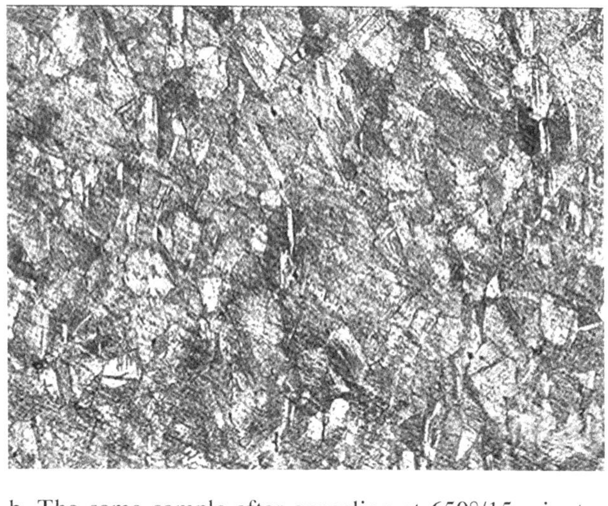

b. The same sample after annealing at 650°/15 minutes showing complete recrystallisation. The slip lines were due to a further reduction after the annealing

Figure 7.6 Photomicrographs of a 30% cold-worked 10% Sn bronze after a cycle of rolling and annealing

7.6 Microstructures of the cold-worked bronzes

The estimate of reduction (in thickness) rates by cold-working in archaeological bronzes is often based on the degree of elongation of sulphide inclusions (Northover 1996, 323; Baboula and Northover 1999, 148). This present experimental series provides many photomicrographs of bronzes of various composition, cast in different conditions and cold-worked to differing degrees, which may help the archaeometallurgists to estimate the degree of cold-working on ancient objects. A complete series of photomicrographs of the cold-worked samples discussed in this chapter is presented in appendix 5 on the CD accompanying this volume. Some of them (figs. 7.7-7.10) are shown here for discussion.

The density of strain or slip lines of worked bronzes is commonly much higher on the surface than in the centre, as illustrated in figure 7.7a and b on a sample of a 6% tin bronze, cast in sand and air-cooled, which has been subjected to a 10% reduction. The elongation of the grains was not apparent until the reduction reached 30%. At a reduction of 30%, the grain elongation can clearly be seen on the surface (fig. 7.7c) but not in the centre, where a remnant dendritic structure is still visible (fig. 7.7d). At a reduction of 45% to 60%, the original dendritic structure has almost disappeared (fig. 7.7e-h). At a reduction of 80% and above, only aligned bands can be seen (fig. 7.7i-l).

Not all slip lines, however, especially those on the surface, are necessarily a result of cold-working. The experiments reported here have demonstrated that surface finishing, e.g. sand-gritting, can also produce slip lines on the surface (fig. 7.8). Manufacturing and finishing processes therefore have to be taken into consideration when ancient bronzes are being examined.

Estimates of the degree of reduction in thickness from sulphide inclusions alone should be treated with caution because sulphide inclusions can be different shapes (Wang Quanyu 2002, 233), which may resist deformation to different degrees. In figure 7.9 differential elongation of the sulphide inclusions can be seen, which could be due to a difference in their original shapes, e.g. the one on the upper left corner could have been thinner to start with than the other two on the right side of the figure. In contrast to sulphide inclusions, deformation of the $\alpha+\delta$ eutectoids is sometimes very slight, even in a sample which has undergone a 60% reduction (fig. 7.10).

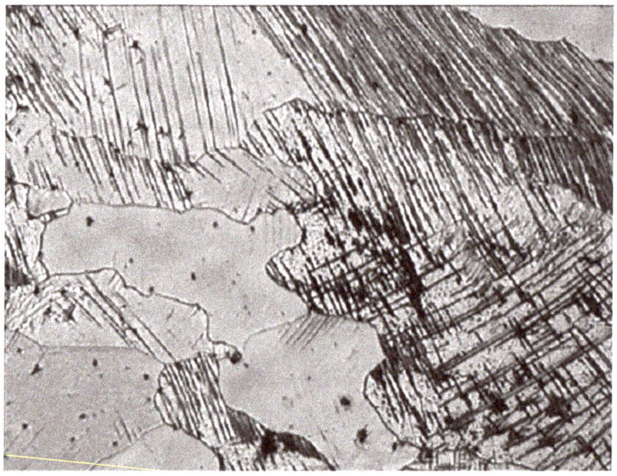

a. 10% reduction, surface of sample (image width 0.65 mm)

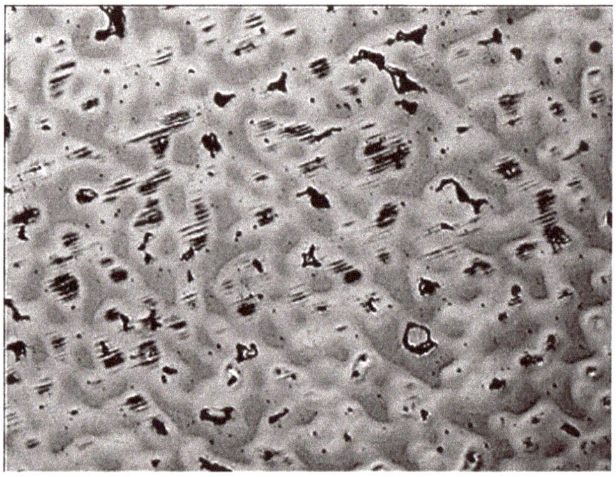

b. 10% reduction, centre of sample (image width 0.65 mm)

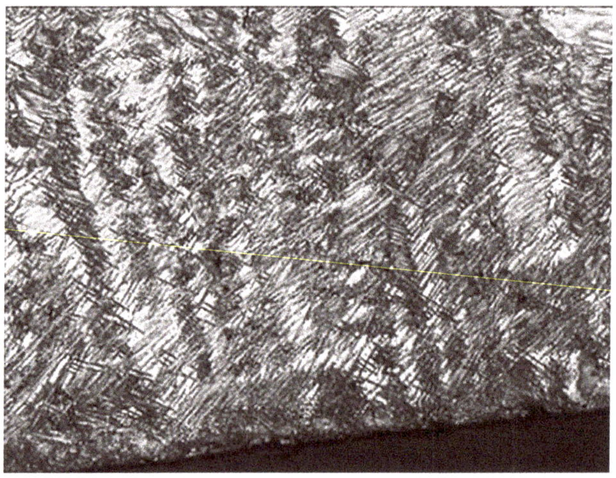

c. 30% reduction, surface of sample (image width 0.65 mm)

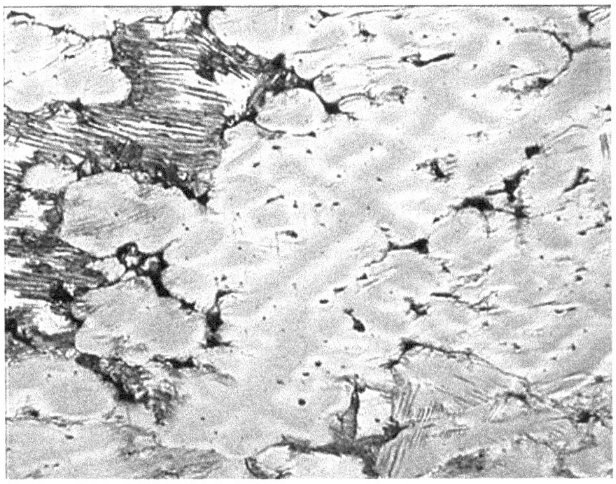

d. 30% reduction, centre of sample (image width 0.65 mm)

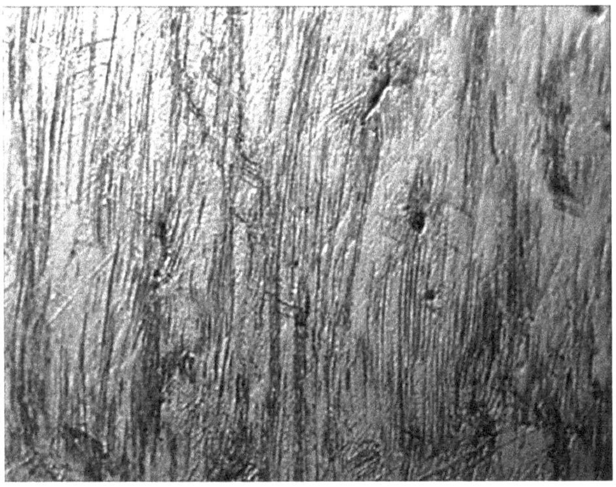

e. 45% reduction, surface of sample (image width 0.13 mm)

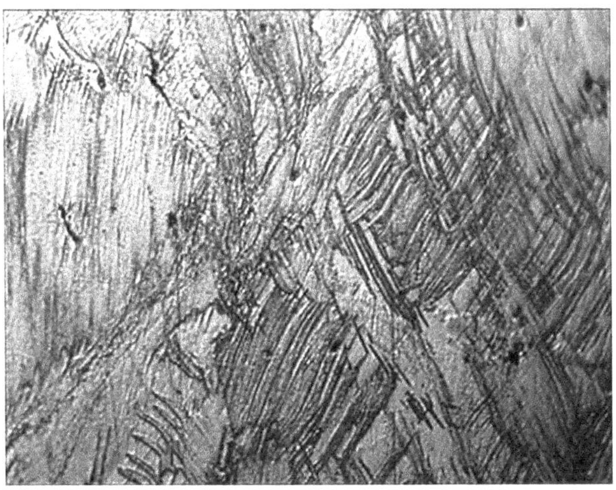

f. 45% reduction, centre of sample (image width 0.13 mm)

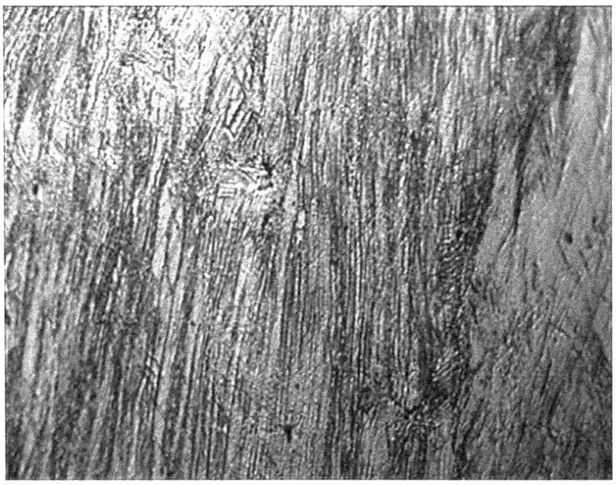

g. 60% reduction, surface of sample (image width 0.13 mm)

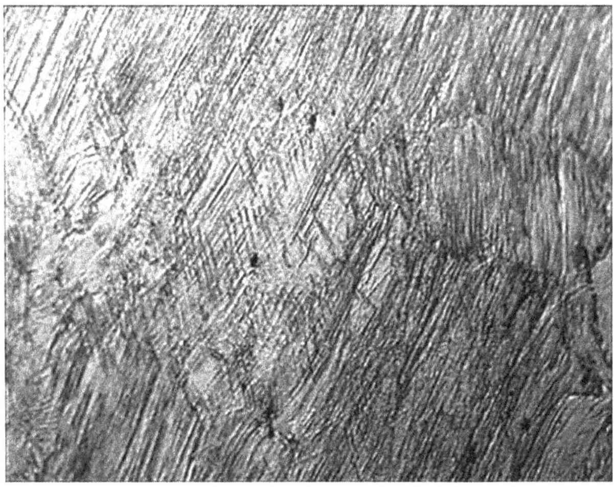

h. 60% reduction, centre of sample (image width 0.13 mm)

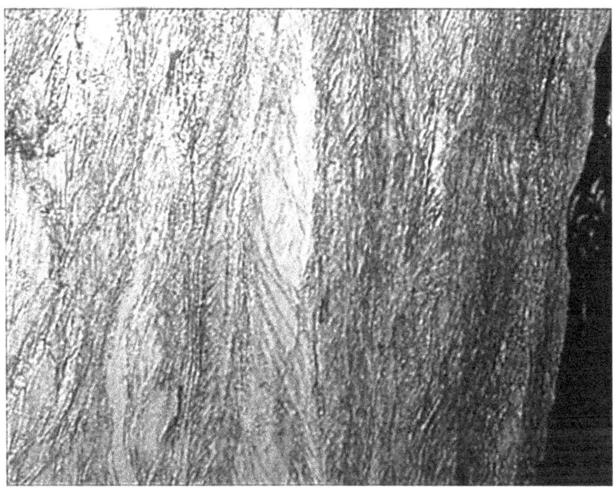

i. 80% reduction, surface of sample (image width 0.13 mm)

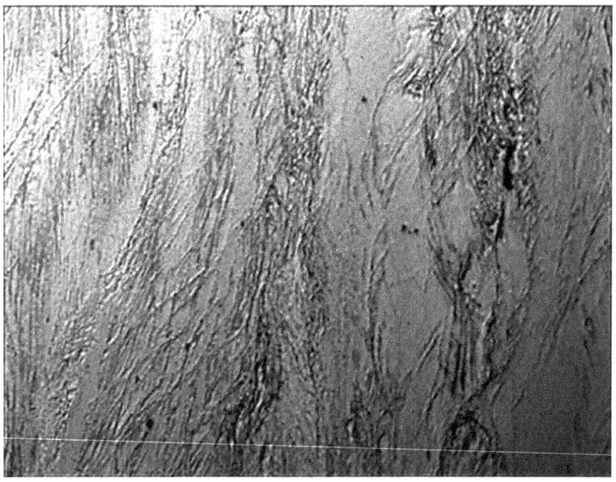

j. 80% reduction, centre of sample (image width 0.13 mm)

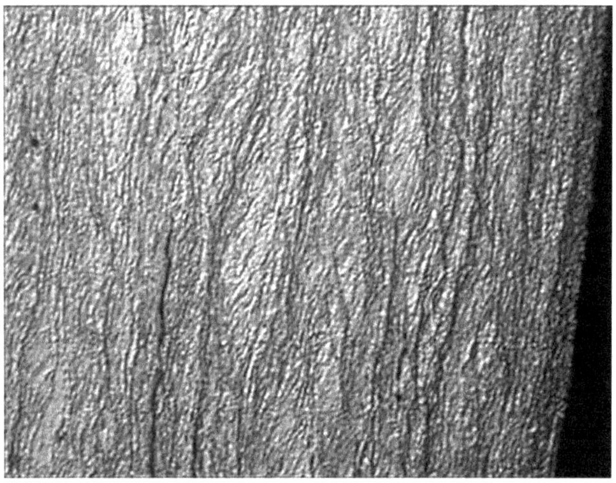

k. 90% reduction, surface of sample (image width 65 μm)

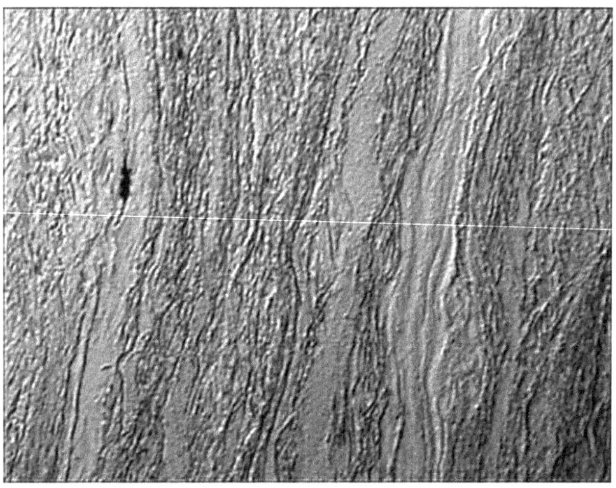

l 90% reduction, centre of sample (image width 65 μm)

Figure 7.7 Photomicrographs of 6% Sn bronze (cast in sand and air-cooled) at various reduction rates by cold-rolling without annealing

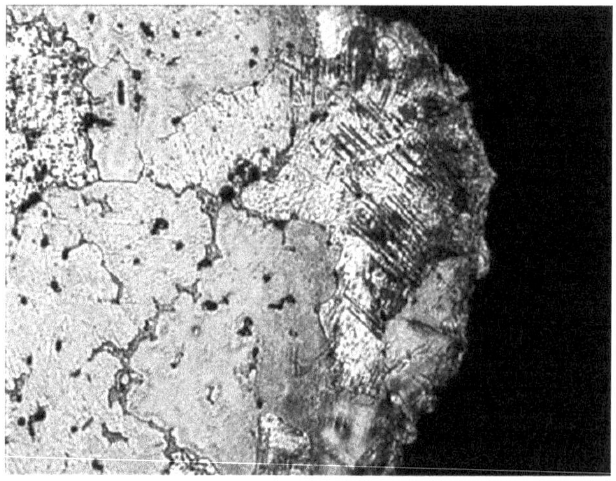

Figure 7.8 Photomicrograph of bronze containing 10% Sn and 2% Pb (cast in sand and air-cooled), showing slip lines on the surface caused by sand gritting employed for cleaning the surface (image width 0.65 mm)

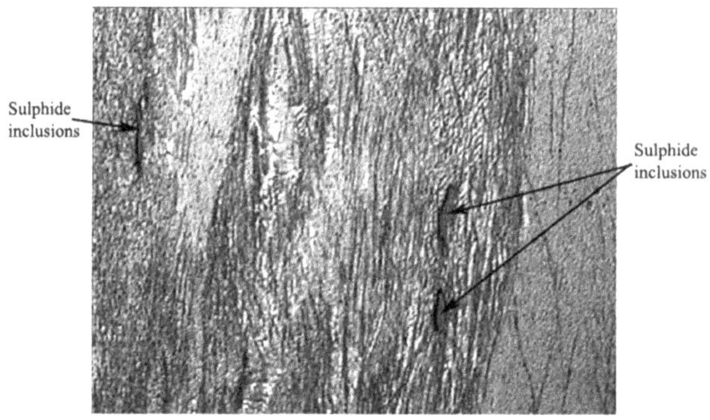

Figure 7.9 Photomicrograph of cold-worked bronze with 60% reduction containing 6% Sn (cast in clay and air-cooled), showing elongation of sulphide inclusions (image width 0.13 mm)

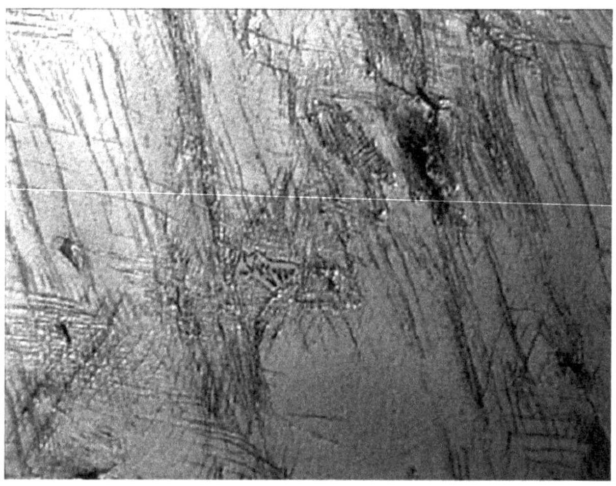

Figure 7.10 Photomicrograph of cold-worked bronze with 60% reduction containing 6% Sn (cast in clay and air-cooled), showing unaffected $\alpha+\delta$ eutectoids (image width 0.13 mm)

7.7 Conclusions

The experiments show that water-quenched bronzes can stand greater reduction by cold-working than air-cooled bronzes can. However, the cooling method did not affect the hardness of the bronzes, and there is no clear correlation between hardness and moulding material. It does seem that the bronzes from the sand moulds can stand greater reduction than those cast in clay and bronze moulds.

Alloying composition, such as copper with tin or lead or a combination of the two, has a major effect on the workability of the bronzes. The lower the tin content, the greater the reduction that can be achieved; bronzes containing 15% tin cannot be reduced in thickness by cold-working beyond 30%. It seems that the addition of lead can also affect the workability. Cracking appeared earlier in those bronzes containing 6% tin with varying amounts of lead than in bronzes containing 6% tin alone.

Deformation, e.g. by cold-working prior to annealing, is a major factor in recrystallisation; the higher the deformation, the more complete the recrystallisation. This is well known in the metallurgical literature and has been shown to apply to the bronzes produced in this series.

The microstructure of a cold-worked bronze can be very different at the surface from that in the centre in terms of the density of slip lines, even for relatively small objects such as the bronze axes cast in this series of experiments. The elongation and appearance of sulphide inclusions affected by cold-working may be uneven because of the difference in their original shapes. All these factors have to be taken into account and a cautious approach to estimating the degree of cold-working on archaeological bronzes has to be adopted

8 Behaviour of the alloying elements tin and lead during melting and casting

8.1 Introduction

Compositional analyses of the experimentally cast bronzes were carried out by Sheffield Analytical Service using an ICP-OES Perkin Elmer 3300 Radial. The parameters for the analyses were as follows: sample weight 100 mg, accuracy 1% of content, and detection limit 0.02%. Two samples, one from the edge and one from the section in the middle of each object, were taken for analysis to counteract possible problems of segregation and to obtain an average composition.

8.2 Results and discussion

The ICP-OES results, along with the composition of the original charge as weighed in, using an electronic balance with a precision of 0.1 g, for each batch of metal, are listed in table 8.1. Since on average two to three moulds were filled with a single batch of molten metal of the same composition, not all sample numbers are present in table 8.1. Furthermore, the table lists the batches in order of alloying composition and hence the batch/sample numbers in table 8.1 are not sequential. A complete list of sample numbers for each batch can be found in chapter 3 (table 3.1), chapter 4 (table 4.3) and chapter 5 (table 5.1).

Table 8.1 *Composition of bronzes cast in different moulds analysed by ICP-OES and original charges as weighed in for each batch of metal (all figures in wt %)*

	Sand moulds				**Clay moulds**				**Bronze moulds**			
	Batch/ sample no.	**Cu**	**Sn**	**Pb**	**Batch/ sample no.**	**Cu**	**Sn**	**Pb**	**Batch/ sample no.**	**Cu**	**Sn**	**Pb**
	42				112				204			
Edge[1]		97.79	2.18	0.02		98.10	1.88	0.02		98.10	1.94	0.02
Flat[2]		97.77	2.19	0.02		98.29	1.69	0.02		98.10	1.88	0.02
Mean		97.78	2.18	0.02		98.19	1.79	0.02		98.10	1.91	0.02
Original charge		97.95	2.05	0.00		98.00	2.00	0.00		98.01	1.99	0.00
	35				114				206			
Edge		94.10	5.92	0.01		94.04	5.96	0.01		94.10	5.93	0.01
Flat		94.10	5.90	0.02		93.84	6.12	0.02		94.70	5.29	0.02
Mean		94.10	5.91	0.01		93.94	6.04	0.02		94.40	5.61	0.02
Original charge		94.01	5.99	0.00		94.00	6.00	0.00		93.99	6.01	0.00
	44				118				208			
Edge		89.23	10.72	0.02		90.55	9.42	0.01		90.30	9.75	0.01
Flat		89.28	10.68	0.01		89.96	10.02	0.01		90.00	10.03	0.01
Mean		89.26	10.70	0.01		90.26	9.72	0.01		90.15	9.89	0.01
Original charge		89.99	10.01	0.00		90.00	10.00	0.00		89.99	10.01	0.00
	38				121				210			
Edge		84.61	15.34	0.00		85.11	14.83	0.01		85.10	14.93	0.00
Flat		84.86	15.07	0.00		84.92	15.07	0.00		84.80	15.20	0.00
Mean		84.73	15.20	0.00		85.02	14.95	0.00		84.95	15.06	0.00
Original charge		84.97	15.03	0.00		85.00	15.00	0.00		85.01	14.99	0.00
	40				124				212			
Edge		76.92	23.02	0.00		76.90	23.09	0.01		76.70	23.26	0.00
Flat		76.78	23.16	0.00		76.09	23.90	0.01		76.40	23.55	0.00
Mean		76.85	23.09	0.00		76.49	23.49	0.01		76.55	23.41	0.00
Original charge		77.06	22.94	0.00		77.01	22.99	0.00		77.04	22.96	0.00
	46				127				214			
Edge		95.96	1.97	2.04		96.41	1.76	1.82		96.20	1.87	1.90
Flat		96.20	1.84	1.94		96.27	1.78	1.95		96.10	1.92	1.97
Mean		96.08	1.91	1.99		96.34	1.77	1.89		96.15	1.90	1.93
Original charge		96.04	2.00	1.96		96.00	2.01	1.99		96.00	2.02	1.98
	48				130				216			
Edge		91.44	1.90	6.64		92.64	1.81	5.56		92.20	1.88	5.95

Table 8.1 ***(contd.) Composition of bronzes cast in different moulds analysed by ICP-OES and original charges as weighed in for each batch of metal (all figures in wt %)***

Flat		91.49	1.94	6.56		92.64	1.72	5.63		92.30	1.86	5.87
Mean		91.47	1.92	6.60		92.64	1.77	5.60		92.25	1.87	5.91
Original charge		91.97	2.01	6.02		92.00	2.01	5.99		91.97	2.04	5.99
	50				133				218			
Edge		87.51	1.94	10.55		89.85	1.80	8.34		88.30	1.83	9.92
Flat		87.54	1.91	10.52		89.61	1.90	8.49		88.10	1.92	9.95
Mean		87.53	1.93	10.54		89.73	1.85	8.42		88.20	1.88	9.94
Original charge		88.04	1.96	10.00		88.10	1.98	9.92		88.00	2.00	10.00
	52				136				220			
Edge		91.85	6.14	2.01		92.89	5.41	1.70		92.40	5.61	2.02
Flat		91.75	6.18	2.05		92.34	5.77	1.89		92.20	5.78	2.01
Mean		91.80	6.16	2.03		92.57	5.59	1.80		92.30	5.70	2.02
Original charge		91.85	6.01	2.14		92.02	6.00	1.98		91.99	6.00	2.01
	54				139				222			
Edge		88.40	5.39	6.16		88.24	5.84	5.93		88.30	5.80	5.94
Flat		88.13	5.52	6.35		88.24	5.76	5.99		88.40	5.74	5.82
Mean		88.26	5.46	6.26		88.24	5.80	5.96		88.35	5.77	5.88
Original charge		88.50	5.50	6.00		87.83	6.09	6.08		88.00	6.00	6.00
	56				142				224			
Edge		87.60	6.27	6.18		84.95	5.49	9.57		84.00	5.92	10.13
Flat		87.60	6.21	6.17		82.21	5.74	10.04		84.00	5.95	10.09
Mean		87.60	6.24	6.18		84.58	5.62	9.81		84.00	5.94	10.11
Original charge		85.99	6.01	10.00		85.99	5.99	10.02		85.99	6.03	9.98
	58				145				226			
Edge		87.92	9.99	2.08		88.65	9.48	1.86		88.40	9.61	2.03
Flat		87.90	10.01	2.08		88.25	9.75	1.96		88.30	9.63	2.03
Mean		87.91	10.00	2.08		88.45	9.62	1.91		88.35	9.62	2.03
Original charge		88.00	10.00	2.00		89.02	9.98	2.00		87.98	10.02	2.00

Note:[1] Edge = sample taken from edge of bronze axe for ICP-OES analysis.
[2] Flat = sample taken from section of bronze axe for ICP-OES analysis.

It can be seen from table 8.1 that there is little difference in the tin and lead content of samples taken from the edge and the section. This is compatible with the microstructures, which indicate that there is little segregation. Table 8.1 also shows that the actual composition, as measured by ICP-OES, is in most cases pretty close to the nominal composition, i.e. as weighed in. This is demonstrated in figure 8.1, which compares the weighed-in (nominal) and ICP-OES (actual) values of tin. Figure 8.2 compares nominal and actual values of lead and it can be seen that, again, there is good agreement overall except in one sample (no. 133), which shows a notable discrepancy between the two lead values.

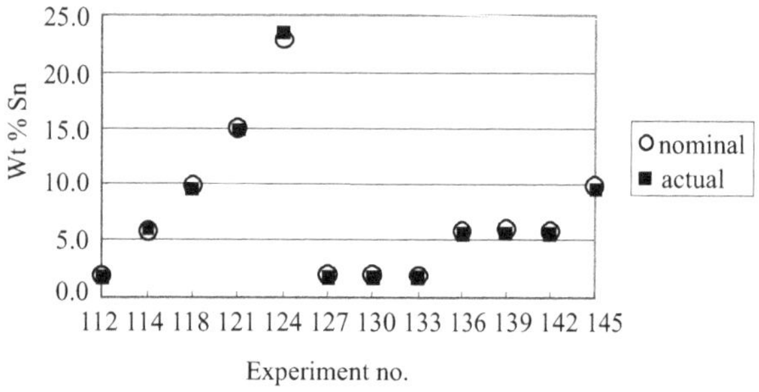

Figure 8.1 Comparison of the nominal tin content as weighed in to the charge and actual tin values of the bronzes cast in clay moulds and analysed by ICP-OES. The experiment numbers refer to the batch/sample numbers in table 8.1

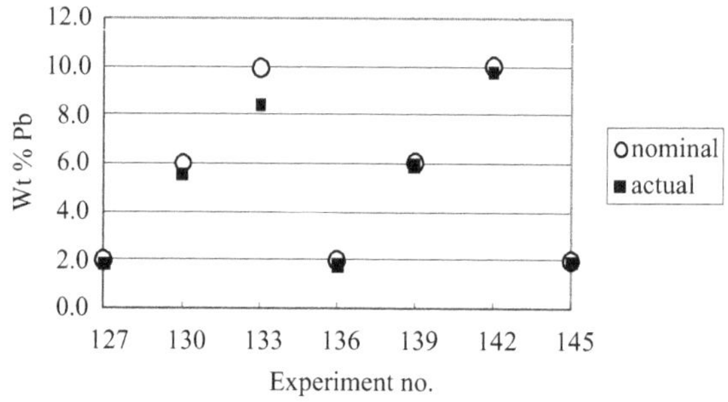

Figure 8.2 Comparison of the nominal lead content as weighed in to the charge and actual lead values of the bronzes cast in clay moulds and analysed by ICP-OES. The experiment numbers refer to the batch/sample numbers in table 8.1

It has been reported that loss of volatile alloying elements such as arsenic can occur during melting, and that the loss is more severe in oxidising than in reducing conditions (McKerrell and Tylecote 1972, 212). Davies and Oelmann (1985, 76) also point out that considerable loss of metals with a low boiling point, such as zinc, can occur during melting in air, i.e. in oxidising conditions. Experimental casting carried out by Northover (pers. comm.) and his team in Oxford also suggest a loss of lead during melting.

The percentage of weight loss, or gain, of tin and lead for each batch in the series of experiments is listed in tables 8.2 and 8.3, respectively. The loss in these tables and in figures 8.3–8.6 is represented by positive values and the gain by negative values.

Table 8.2 ***Content of tin (wt %) in the charges and in the actual casts*** [1]

	Batch/sample no.											
	42	**35**	**44**	**38**	**40**	**46**	**48**	**50**	**52**	**54**	**56**	**58**
Sand moulds												
Sn% of charge	2.05	5.99	10.01	15.03	22.94	2.00	2.01	1.96	6.01	5.50	6.01	10.00
Sn% of cast object	2.18	5.91	10.70	15.20	23.09	1.91	1.92	1.93	6.16	5.46	6.24	10.00
Loss/gain[1] of Sn (%)	-6.34	1.34	-6.89	-1.13	-0.65	4.50	4.48	1.53	-2.50	0.73	-3.83	0.00
	112	**114**	**118**	**121**	**124**	**127**	**130**	**133**	**136**	**139**	**142**	**145**
Clay moulds												
Sn% of charge	2.00	6.00	10.00	15.00	22.99	2.01	2.01	1.98	6.00	6.09	5.99	9.98
Sn% of cast object	1.79	6.04	9.72	14.95	23.49	1.77	1.77	1.85	5.59	5.80	5.62	9.62
Loss/gain of Sn (%)	10.50	-0.67	2.80	0.33	-2.17	11.94	11.94	6.57	6.83	4.76	6.18	3.61
	204	**206**	**208**	**210**	**212**	**214**	**216**	**218**	**220**	**222**	**224**	**226**
Bronze moulds												
Sn% of charge	1.99	6.01	10.01	14.99	22.96	2.02	2.04	2.00	6.00	6.00	6.03	10.02
Sn% of cast object	1.91	5.61	9.89	15.06	23.41	1.90	1.87	1.88	5.70	5.77	5.94	9.62
Loss/gain of Sn (%)	4.02	6.66	1.20	-0.47	-1.96	5.94	8.33	6.00	5.00	3.83	1.49	3.99

Note:[1] Loss is represented by positive, gain by negative values.

Table 8.3 ***Content of lead (wt %) in charge and actual cast*** [1]

	Batch/sample no.						
	46	**48**	**50**	**52**	**54**	**56**	**58**
Sand moulds							
Pb% of charge	1.96	6.02	10.00	2.14	6.00	10.00	2.00
Pb% of cast object	1.99	6.60	10.54	2.03	6.26	6.18	2.08
Loss/gain of Pb (%)	-1.53	-9.63	-5.40	5.14	-4.33	38.20	-4.00
	127	**130**	**133**	**136**	**139**	**142**	**145**
Clay moulds							
Pb% of charge	1.99	5.99	9.92	1.98	6.08	10.02	2.00
Pb% of cast object	1.89	5.60	8.42	1.80	5.96	9.81	1.91
Loss/gain of Pb (%)	5.03	6.51	15.12	9.09	1.97	2.10	4.50
	214	**216**	**218**	**220**	**222**	**224**	**226**
Bronze moulds							
Pb% of charge	1.98	5.99	10.00	2.01	6.00	9.98	2.00
Pb% of cast object	1.93	5.91	9.94	2.02	5.88	10.11	2.03
Loss/gain of Pb(%)	2.53	1.34	0.60	-0.50	2.00	-1.30	-1.50

Note:[1] Loss is represented by positive, gain by negative values.

When the loss, or gain, of tin is plotted against the type of mould (fig. 8.3), it can be seen that there is considerably more loss of tin (values > 0) in the clay and bronze moulds than in the sand moulds. The same can be observed for the lead values (fig. 8.4). The loss of tin seems to be exacerbated by the addition of lead, in particular for 2% and 10% tin bronzes cast in sand and bronze moulds (see fig. 8.3). On the other hand, the addition of lead to 6% tin bronzes seems to reduce the loss of tin. This complex behaviour of alloying elements requires further detailed studies which lie beyond the scope of this research project.

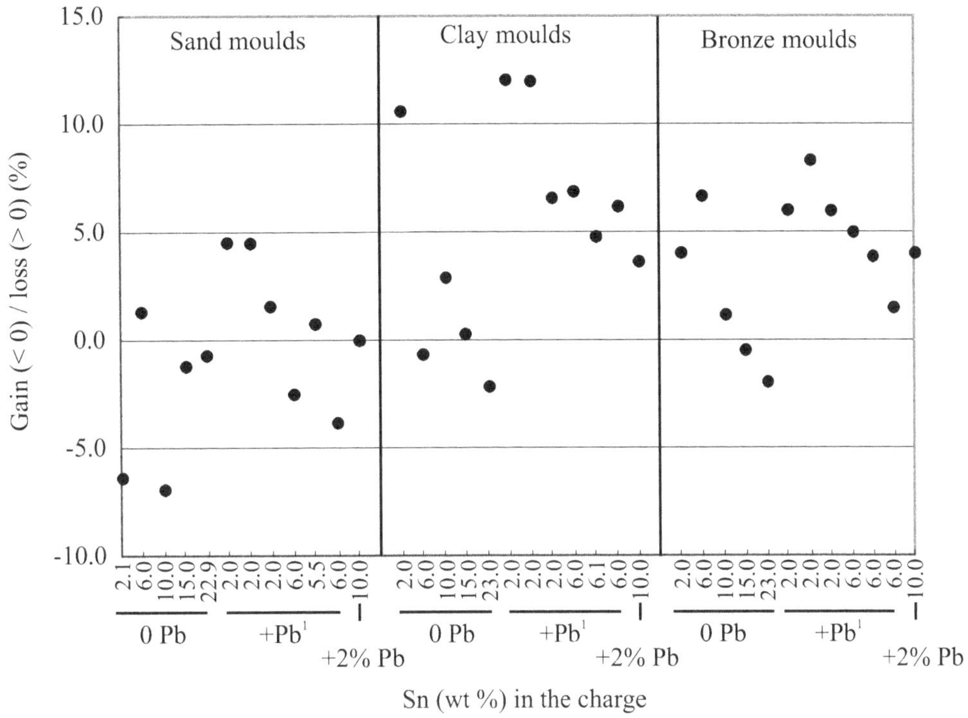

Figure 8.3 Loss or gain of tin in the three moulding materials

Note:[1] The lead contents in the alloys of 2 and 6% Sn are 2, 6 and 10% from left to right (see table 8.1)

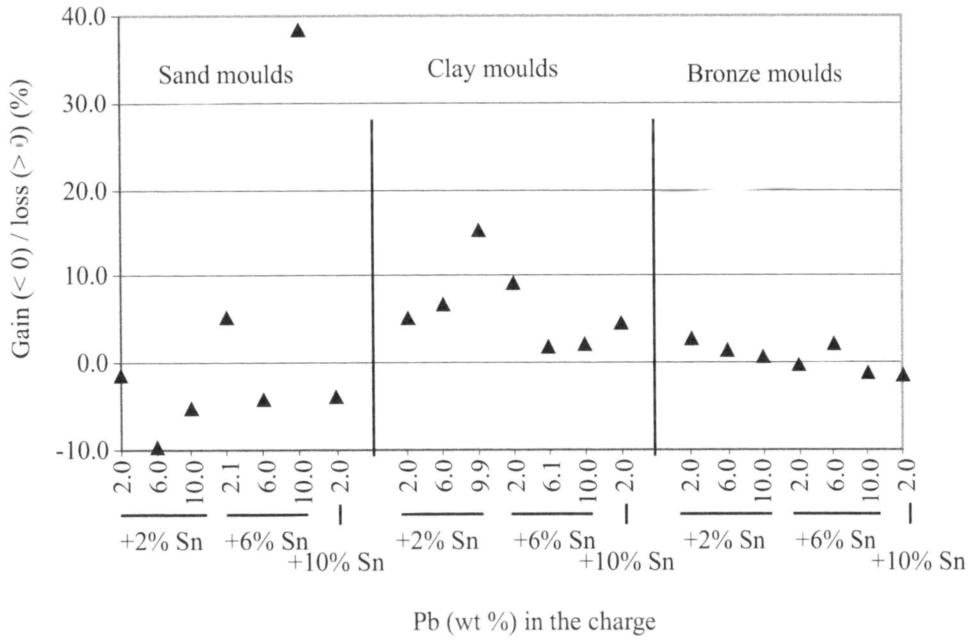

Figure 8.4 Loss or gain of lead in the three moulding materials

The loss of lead in our experiments is below 15% for all bronzes but one, the 6% tin + 10% lead cast in a sand mould. As discussed in chapter 2.2, tin and lead were added to the molten copper just before pouring to avoid loss of alloying metals. Charcoal was put on both the top and bottom of the crucible during melting to provide a reducing atmosphere and this probably explains the low rate of loss of tin and lead in the experimental series discussed here.

It is more difficult to explain the gain of tin and lead in some of the casts. This could be due to a loss of copper during melting as copper sometimes stuck to the stirring rod and could not be put back into the melt. The loss, or gain, of both tin and lead fall within 10% in 31 out of 36 batches of metal used for our casting experiments.

When the loss, or gain, of tin is plotted regardless of the moulding material used (fig. 8.5), it can be seen that the loss of tin is more severe in bronzes with a lower tin content, particularly in the 2 and 6% tin bronzes, than in those with high tin values.

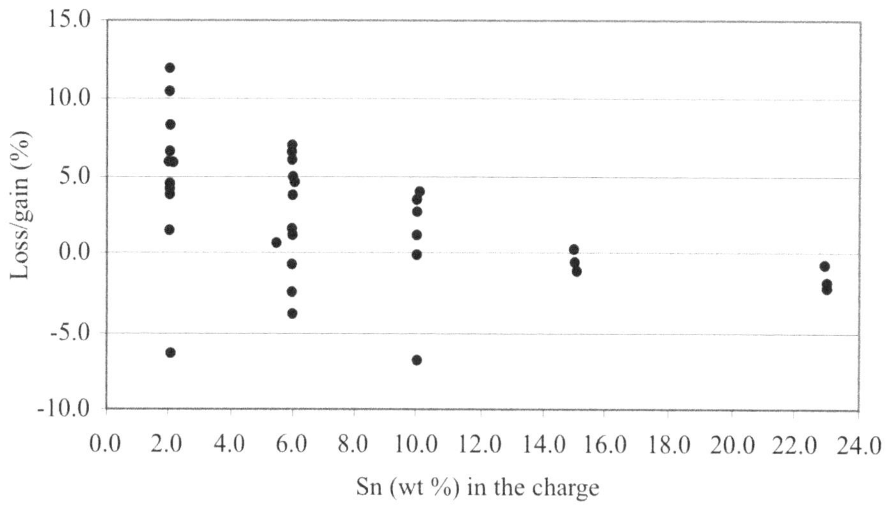

Figure 8.5 Overall view of loss (> 0) or gain (< 0) of tin regardless of moulding material

Conversely, plotting the loss, or gain, of lead regardless of the type of mould (fig. 8.6) shows that the higher the lead content, the more severe the loss.

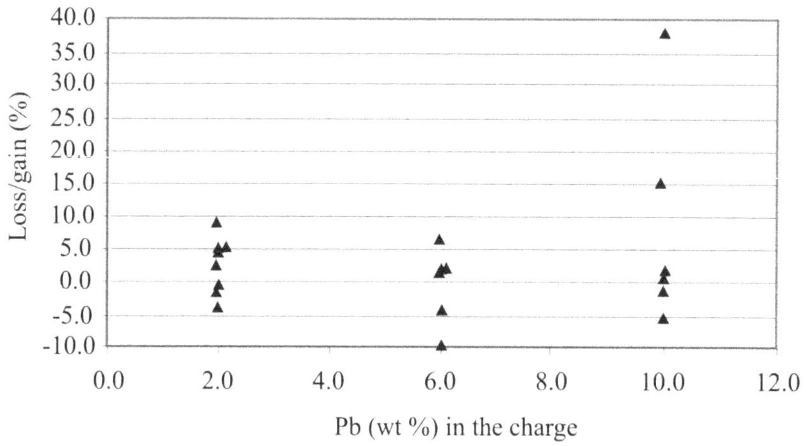

Figure 8.6 Overall view of loss (> 0) or gain (< 0) of lead regardless of moulding material

8.3 Conclusions

Comparison of nominal quantities of alloying elements added to the charge and actual values of the alloying elements in the cast bronzes analysed by ICP-OES established that the compositions are in reasonable agreement for most of the metal batches used in the series of experimental castings reported here. The weight loss, or gain, of both tin and lead falls within 10% in 31 out of the 36 batches of metal. It seems that there is a greater loss of tin and of lead in the clay and bronze moulds than in the sand moulds. The loss of tin was more severe in the casts with a low tin content whereas the loss of lead was more severe in alloys with a high lead content. The addition of lead to tin bronzes seems to exacerbate the loss of tin in most, but not all, cases. This complex behaviour and interrelation of alloying elements requires further study.

9 Conclusions and suggestions for future research

9.1 Conclusions

The aim of this project was to build up a much-needed reference collection of the microstructure and mechanical behaviour of copper-based alloys known to have been used in prehistory. Modern metallurgical researchers are usually not interested in these alloys because they have little or no relevance to material used now. Yet such a database is vital for archaeologists if they are to understand ancient production processes of cast bronze artefacts. For the purpose of building up such a reference collection, a series of casting experiments was conducted under controlled conditions aiming, as closely as possible within modern foundry constraints, to replicate prehistoric casting conditions. This meant avoiding additions to the metal during the process of melting such as the fluxes and deoxidants that are commonly used in modern foundry work to produce flawless casts.

Although a very experienced foundry technician was participating in the casting experiments, it took considerable time, effort and preliminary casting experiments to produce the series of non-porous casts in moulds of sand, clay and bronze.

Flat axes, characteristic artefacts of the Chalcolithic and Early Bronze Age in Europe, were cast in moulds of sand, clay and bronze using different cooling regimes. These included air-cooling and water-quenching for the bronzes cast in sand moulds. For the clay moulds three different cooling methods were employed: the molten metal was poured into (a) moulds preheated to 350°C and left to cool in air, (b) unpreheated moulds and cooled in air and (c) unpreheated moulds and quenched in water. For casting in bronze moulds, water-quenching could not be considered because of safety regulations, so that the bronze moulds were either preheated to 350°C or flame-warmed. The metal used for the experiments included a wide range of tin and leaded tin alloys.

The microstructure, dendritic arm spacing, grain size and microhardness of each cast were studied, as was the malleability of the metals after cold-working and annealing. The composition of each batch of metal was analysed by ICP-OES and the behaviour of alloying elements during casting was investigated. All the results have been presented in the form of tables and figures and a complete set of colour photomicrographs is given on the accompanying CD supplementing the black and white photomicrographs presented in the text.

A study of the microstructure of the ***bronzes cast in sand moulds*** (chapter 3) revealed that the low-tin bronzes, including leaded low-tin bronzes, had a granular microstructure (plate 1) whereas all other bronzes with a higher tin content had dendritic structures. A clear correlation between cooling rates and dendritic arm spacing was found, with water-quenched bronzes having consistently smaller dendritic arm spacing than air-cooled bronzes. Grain size is not so clearly correlated with cooling rate.

Microhardness values were found to be dependent mainly on the composition of the bronzes, although almost all water-quenched casts have higher hardness values than air-cooled bronzes.

Water-quenched bronzes show an appreciably different microstructure from those that have been air-cooled. In the water-quenched 6% tin bronzes the δ phase could be seen but not in the air-cooled ones of the same composition. The β phase was visible in water-quenched bronzes with a tin content of 15% (plate 2) or higher, which is much lower than indicated in published equilibrium diagrams. The occurrence of both β and δ phases in such low-tin bronzes seems to be peculiar to ancient bronzes.

Bronze artefacts cast in clay moulds (chapter 4) were found to have a finer surface than those cast in sand moulds. However, flashing always occurred when clay moulds were used. Thus, while less time was needed to finish the surface of the artefact, more time had to be spent taking off the flashing to prepare the bronze artefact for consumption.

Again, there was an appreciable difference in microstructure between air-cooled and water-quenched bronzes. The microstructures of the casts are complex. The low-tin bronzes have only a granular structure but low-tin leaded bronzes vary: some of them have granular (plate 3) while others have dendritic structures (plate 4). The δ phase was seen in the water-quenched bronzes containing 2% tin which is much lower than suggested by published equilibrium phase diagrams. The β phase began to appear in water-quenched bronzes with 10% tin, which is also much lower than is found in the phase diagrams.

A clear correlation between cooling rate and dendritic arm spacing was found, but not between cooling rate and grain size, although a general trend towards larger grains in preheated and smaller grains in water-quenched bronzes was observed.

Microhardness values were dependent on the composition, although almost all water-quenched casts were found to have higher values than those that were air-cooled, whether cast in preheated or unpreheated moulds. The addition of lead generally leads to a decrease in hardness, but the addition of only 2% lead seems to increase the hardness of water-quenched bronzes.

For safety reasons experiments with ***bronze moulds*** (chapter 5) had only two cooling regimes. Moulds were either preheated to 350°C or flame-warmed and water-quenching had to be abandoned. It was found that the preheated moulds lasted longer than the flame-warmed ones. The former were used nine times and the latter six times before any damage to the mould appeared. The bronzes cast in these bronze moulds had almost no flashing and the surfaces of the casts were very smooth. In terms of ease of finishing off an artefact, bronze moulds are thus superior to either sand or clay moulds.

The microstructure of all bronzes cast in bronze moulds is dendritic and only the core of the 2% tin bronze showed any granular features. In bronzes containing 6% tin or more, a δ phase could be seen. There was little difference in the microstructure of bronzes cast in preheated and those cast in flame-warmed moulds, probably because the difference in temperature of the moulds prior to casting was not very great. Yet, despite this, it was found that bronzes cast in flame-warmed moulds sometimes had a more pronounced dendritic structure and more $\alpha+\delta$ eutectoids than those cast in moulds preheated to 350°C. Heterogeneity in the size of dendrites, i.e. both coarse and fine dendrites, could be observed in one and the same bronze sample (plate 5) and some leaded tin bronzes had coarse but very short dendrites (plate 6).

No clear correlation between cooling rate and dendritic arm spacing or between cooling rate and microhardness values could be observed, probably again due to the negligible difference in the preheating temperatures of the moulds.

The microstructures of all bronzes, cast in the three different moulding materials, were compared (chapter 6). The differences in the microstructure of the bronzes cast in sand and those cast in clay moulds are not very significant. However, the microstructure of the bronzes cast in bronze moulds differs greatly from these, with significantly smaller dendritic arm spacing than is found in bronzes cast in sand and clay moulds. Variations in microstructure have been observed relating to alloying composition and mould material. Most bronzes containing 2% tin and cast in sand and clay moulds have a granular structure, while those cast in bronze moulds have a dendritic structure, at least at the surface. All casts with 6% tin have a dendritic structure, irrespective of mould material or lead content.

There are no $\alpha+\delta$ eutectoids in any of the bronzes with 2% tin, irrespective of either mould material or lead content. They begin to appear in tin bronzes with a tin content of 6% or higher, irrespective of mould material. However, very few or none are present in leaded bronze with 6% tin. This indicates that the addition of lead does have an effect on the microstructure, although it is known that, in theory, lead is immiscible with copper and thus should not affect the microstructure. However, as lead lowers the melting-point of an alloy, possibly changing the casting temperature, the microstructure is affected. This is most noticeable in 2% tin bronzes with lead (cast in sand and clay moulds), where the higher, 2% tin bronze with 10% lead has a granular structure and the lower leaded bronzes have a mixed granular and dendritic microstructure. All bronzes with a tin content of 10% and higher have a dendritic structure with $\alpha+\delta$ eutectoids. The higher the tin content the more $\alpha+\delta$ eutectoids there are.

From the evidence of the dendritic arm spacing, the cooling rates of air-cooled bronzes for the three moulding materials increase from clay to sand to bronze. The fastest cooling rate in the experimental series was found to be in the casts produced in flame-warmed bronze moulds. An even faster cooling rate can be predicted if a cast produced in a bronze mould were to be quenched in water. The slowest cooling rate was found to be in the casts produced in preheated clay moulds and cooled in air in this series.

The microhardness increases greatly with increasing tin content. There is little difference in hardness between casts made in clay and sand moulds. Most casts produced in bronze moulds have slightly, but consistently, higher hardness values than those cast in clay or sand moulds. The highest hardness values of the as-cast bronzes in the series of experiments reported here were found to be in bronzes cast in bronze moulds and air-cooled and the lowest seem to be in bronzes cast in preheated clay moulds and air-cooled. The results are, of course, closely related to the cooling rates discussed above.

To investigate the ***workability of the bronzes*** cast in the different types of moulds (chapter 7), a series of experiments of cold-working followed by annealing was conducted and the behaviour of the metal was investigated at each step in the series. It was found that water-quenched bronzes can generally stand a greater reduction in thickness by cold-working than can air-cooled bronzes. Bronzes cast in sand can stand more cold-working than those cast in other types of mould.

Alloying compositions have a major effect on the workability of bronzes. The lower the tin content, the higher the reduction that can be achieved. The addition of lead also affects the workability. Bronzes containing 6% tin and

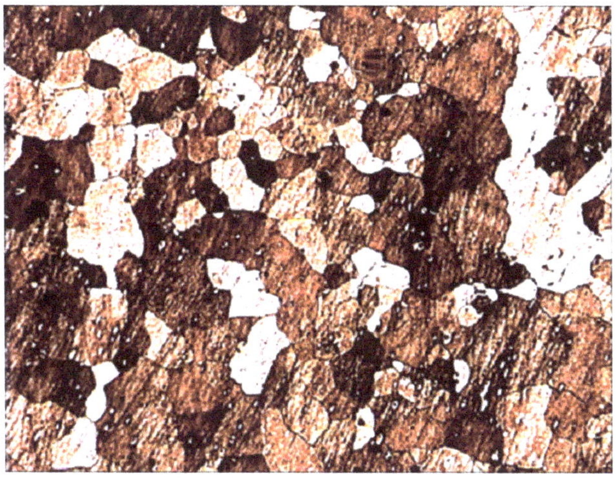

Plate 1 2% Sn bronze, cast in sand mould, air-cooled, showing a granular structure (image width 1.3 mm)

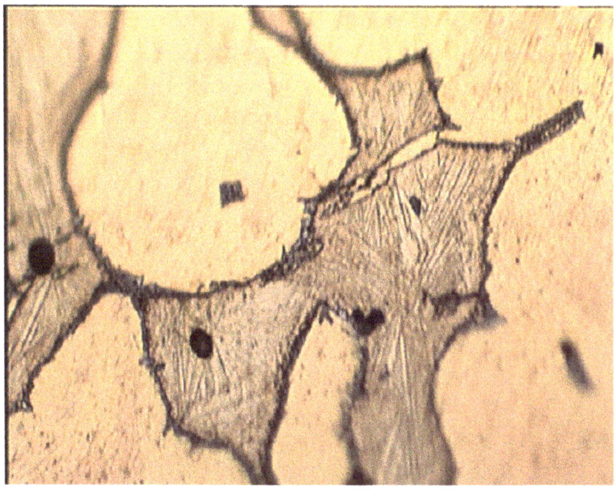

Plate 2 15% Sn bronze, cast in sand mould, water-quenched, showing needle beta phase in the interdendritic regions (image width 0.13mm)

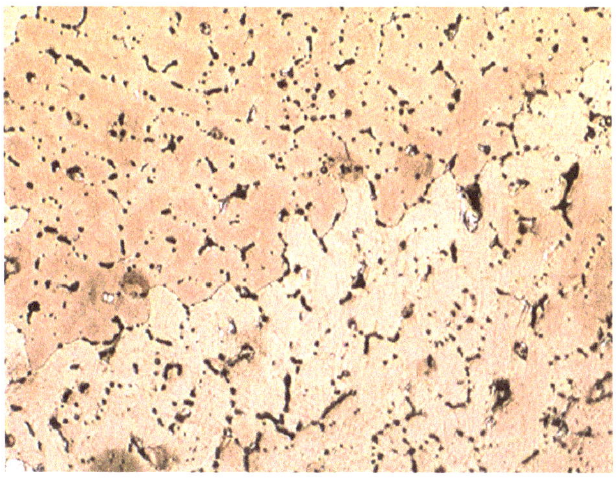

Plate 3 2% Sn + 6% Pb bronze, cast in clay mould, preheated and air-cooled, showing a granular structure (image width 1.3mm)

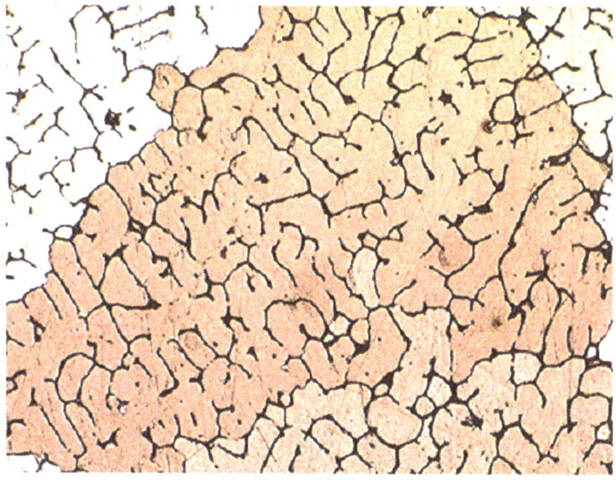

Plate 4 2% Sn +2% Pb bronze, cast in unpreheated clay mould, water-quenched, showing a pronounced dendritic structure (image width 1.3mm).

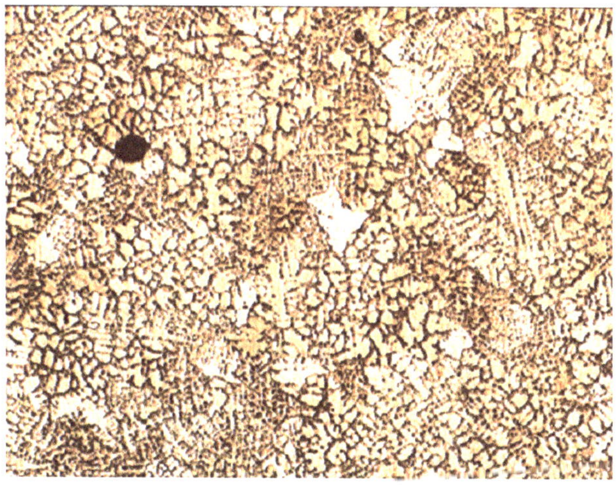

Plate 5 2% Sn+2% Pb bronze, cast in flame-warmed bronze mould and air-cooled, showing both coarse and fine dendrites (image width 1.3mm)

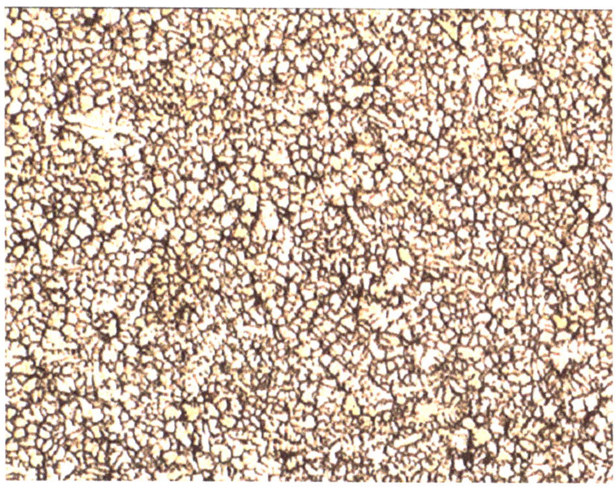

Plate 6 2% Sn+10% Pb bronze, cast in preheated bronze mould and air-cooled, showing very short dendrites (image width 1.3mm)

various amounts of lead were found to crack earlier under cold-working than a 6% tin bronze without lead.

Slip or strain lines are usually taken as an indication of how much working an artefact has undergone. The bronzes in this experimental series have demonstrated that cold-working could affect the microstructure of the centre and the surface quite differently, even for such relatively small artefacts as bronze axes.

All these factors have to be taken into account and a cautious approach to estimating the degree of cold-working in archaeological bronzes should be adopted.

Finally, the ***behaviour of the alloying elements*** tin and lead during melting and casting was studied (chapter 8). A comparison of nominal quantities of alloying elements added to the charge and actual elemental alloying values in the cast bronzes analysed by ICP-OES established that the results are in reasonable agreement for most of the metal batches used in the series of experimental castings reported here. The weight loss, or gain, of both tin and lead fall within 10% in 31 out of the 36 batches of metal used for the castings. It seems that there is a greater loss of tin and of lead from casts made in clay and bronze moulds than from those in sand moulds. The loss of tin was more severe in casts with a low tin content whereas the loss of lead was more severe in alloys with a high lead content. The addition of lead to tin bronzes seems to exacerbate the loss of tin in most but not all cases. This complex behaviour and the interrelationship of the alloying elements require further study.

9.2 Suggestions for further research

Several of the conclusions summarised above support results which are already known from the metallurgical literature. Yet some important factors need to be considered here. Many alloys that were used in prehistory are of no interest to metallurgists and are therefore poorly researched. Moreover, some of the published material, e.g. phase diagrams, was not supported by our results.

A comparison of the different moulding materials showed that bronzes cast in bronze moulds had by far the best surface and least flashing, and thus required least labour to prepare the artefact for consumption. Yet the bronze moulds were troublesome to make and lasted for only nine castings before large cracks appeared. This is in direct contradiction to the longevity of bronze moulds suggested in the literature. Clay moulds were easier to make, even though they needed a lengthy drying and firing regime. They produced bronze artefacts with a good surface but flashing remained a problem and the moulds were not reusable. Sand moulds, though not reusable either, were easiest to make and did not produce too much flashing but the bronze artefact had a rather rough surface.

A comparison of the different ways of producing tin or leaded tin bronze artefacts makes it clear that no one process produces the best bronze. For a bronze that can be well reduced in thickness by cold-working and annealing, e.g. for the production of sheet metal such as was used for making tubular beads and other sheet-metal ornaments, a low tin content and no lead, cast in a sand mould and water-quenched, would yield the optimum material. For maximum hardness, a high-tin bronze cast in a bronze mould and air-cooled would give the best result. It will be up to future archaeometallurgical studies to determine whether such choices were made consciously by prehistoric smiths.

As usual, a study of this kind, though answering many questions, also throws up several new ones. For instance, further research is needed to assess complex behaviour and interrelationships of alloying elements such as loss during melting and casting under conditions prevailing in prehistory. More research is also needed to provide a better assessment of the degree of cold-working on small artefacts, because any assessment based solely on either density of slip lines or shapes of sulphide inclusions is not reliable. The density of slip lines may be very different between the surface and the centre of archaeological artefacts. Since the original shapes of sulphide inclusions can be different, the final shapes after cold-working can also be expected to be different.

We concentrated on three moulding materials in this study, but of course more experiments with stone moulds and also with lost wax moulds should be conducted. Similar studies should also be carried out using simpler furnaces, crucibles and moulding boxes which was not possible in this study as stringent safety regulations in the foundry had to be observed. Finally, more studies with different alloying elements such as arsenic and elements occurring in the Fahlerz ores are needed.

Microstructures and microhardness values for a complete series of alloys cast in three different moulding materials have been presented in this volume. We hope that this, together with the use of compositional results, will greatly aid the reconstruction of the sequence of actions involved in the production processes of ancient bronze artefacts. This in turn will, we hope, stimulate further research into choices made by prehistoric smiths. Set within the context of the society in which the smiths lived, such research could bring us a step closer to understanding how things were made and used in Bronze Age societies.

References

Avner, S. H., 1974. *Introduction to Physical Metallurgy*. London: McGraw-Hill.

Baboula, E. and Northover, P., 1999. Metals technology versus context in Late Minoan burials. In S. M. M. Young, A. M. Pollard, P. Budd and R. A. Ixer (eds.), *Metals in Antiquity*, BAR International Series **792**: 146-52. Oxford: BAR Publishing.

Buchwald, V.F. and Leisner, P., 1990. A metallurgical study of 12 prehistoric bronze objects from Denmark. *Journal of Danish Archaeology* **9**: 64-102.

Budd, P., 1991. Eneolithic arsenical copper: heat treatment and the metallographic interpretation of manufacturing processes. In E. Pernicka and A. Wagner (eds), *Archaeometry '90*. International Symposium on Archaeometry, Heidelberg 1990. Basel: Birkhäuser, 35-44.

Budd, P. and Ottaway, B. S., 1991. The properties of arsenical copper alloys: implications for the development of Eneolithic metallurgy. In P. Budd, B. Chapman, C. Jackson, R. Janaway, B. S. Ottaway (eds.), *Archaeological Sciences 1989*. Oxbow Monograph 9. Oxford: Oxbow, 132-42.

Budd, P. and Ottaway, B. S., 1995. Eneolithic arsenical copper: chance or choice? In: B. Jovanovic (ed.), *Ancient Mining and Metallurgy in Southeast Europe*. Belgrade: Archaeology Institute 1995, 95-102.

Bunk, W.G.J. and Kuhnen, H.P., 2003. Roman time copper ingots found at Trier: original and experiment. In Proceedings of the International Conference *Archaeometallurgy in Europe*. Milan 2003, 393-401.

Campbell, J., 1991. *Castings*. Oxford: Butterworth-Heinemann.

Chadwick, R., 1939. The effect of composition and constitution on the working and on some physical properties of the tin bronzes. *Journal of the Institute of Metals*, **64(1):** 331-46.

Childe, V. G., 1930. *The Bronze Age*. Cambridge: Cambridge University Press.

Coghlan, H. H., 1960. Prehistoric working of bronze and arsenical copper. *Sibrium*, **5**: 145-52.

1968. A note on prehistoric casting moulds. *Bulletin of the Historical Metallurgy Group*, **2(2)**: 73.

Coghlan, H. H. and Raftery, J., 1961. Irish prehistoric casting moulds. *Sibrium*, **6**: 223-44.

Collins, A. E. P., 1970. Bronze Age moulds in Ulster. *Ulster Journal of Archaeology*, **33**: 23-36.

Davies, D. J. and Oelmann, L. A., 1985. *Metallurgical Processes and Production Technology*. London: Pitman.

Eccleston, M. and Ottaway, B. S., 2002. Experimental casting of copper and bronzes in sand moulds. In E. Jerem and K. T. Biro (eds.), *Archaeometry '98*, BAR International Series **792**: 185-89. Oxford: BAR Publishing.

Faoláin, N. Ó. and Northover J. P., 1998. The technology of Late Bronze Age sword production in Ireland. *The Journal of Irish Archaeology*, **9**: 69-88.

Fasnacht, W., 2001. Zehn Jahre Kupfer- und Bronzeguss im Experiment-oder die wiederholte Erfindung der Rades. *Zeitschrift für Schweizerische Archäologie und Kunstgeschichte* **58**, 67-72.

Gettens, R. J., 1969. *The Freer Chinese Bronzes*, vol. II, *Technical Studies*. Washington: Smithsonian Institution.

Goodway, M. and Conklin, H., 1987. Quenched high-tin bronzes from the Philippines. *Archaeomaterials*, **2(1)**: 21-7.

Hanson, D. and Marryat, C. B., 1927. Effects of arsenic on copper. *Journal of the Institute of Metals*, **37(1):** 121-43.

Hodges, H. W. M., 1958-9. The Bronze Age moulds of the British Isles, part 1: Scotland and northern England – moulds of stone and clay. *Sibrium*, **4**: 129-37.

1960. The Bronze Age moulds of the British Isles, part 2: England and Wales - moulds of stone and bronze. *Sibrium*, **5**: 153-62.

Hua Jueming, 1985. *Shi Jie Ye Jin Fa Zhan Shi* (A history of world metallurgy). Beijing: Science and Technology Literature Publishing House.

Junghans, S., Sangmeister, E. and Schröder, M., 1960. *Kupfer und Bronze in der Frühen Metallzeit Europas*. Berlin: Gebr. Mann.

- 1968. *Kupfer und Bronze in der Frühen Metallzeit Europas*. Berlin: Gebr. Mann.

- 1974. *Kupfer und Bronze in der Frühen Metallzeit Europas*. Berlin: Gebr. Mann.

Junk, M., 2003. Material properties of copper alloys containing arsenic, antimony, and bismuth. The material of Early Bronze Age ingot torques. PhD Dissertation: Bergakademie Freiberg. PDF-Version 2003.

Lamm, K., 1973. The manufacture of jewellery during the Migration Period at Helgö in Sweden. *Bulletin of the Historical Metallurgy Group*, **7(2)**: 1-7.

Leahy, K., 1977. A survey of casting moulds from the Middle and Late Bronze Age in Great Britain and Ireland. Unpublished undergraduate dissertation: University of Leicester.

Lechtman, H., 1996. Arsenic bronze: dirty copper or chosen alloy? A view from the Americas. *Journal of Field Archaeology*, **23**: 477-514.

Lechtman, H. and Klein, S., 1999. The production of copper-arsenic alloys (arsenic bronze) by cosmelting: modern experiment, ancient practice. *Journal of Archaeological Science*, **26**: 497-526.

McKerrell, H. and Tylecote, R. F., 1972. The working of copper-arsenic alloys in the Early Bronze Age and the effect on the determination of provenance. *Proceedings of the Prehistoric Society*, **30**: 209-18.

Matsuda, T., 1928. On the quenching and tempering of brass, bronze, and aluminum-bronze. *Journal of the Institute of Metals*, **39(1)**: 67-108.

Meeks, N. D., 1993. Patination phenomena on Roman and Chinese high-tin bronze mirrors and other artefacts. In S. La Niece and P. Craddock (eds.), *Metal Plating and Patination*, 63- 84. Oxford: Butterworth-Heinemann.

Mohen, J. P., 1978. Bronze moulds of the Bronze Age. *Antiquities Nationales*, **10**: 23-32.

Needham, S., 1980. An assemblage of Late Bronze Age metalworking debris from Dainton, Devon. *Proceedings of the Prehistoric Society*, **46**: 177-215.

Northover, J. P. 1989. Properties and use of arsenic copper alloys. In A. Hauptmann, E. Pernicka and G. A. Wagner, (eds.) *Old World Archaeometallurgy*. Proceedings of the International Symposium, Heidelberg 1987. Der Anschnitt Beiheft **7**: 111-18.

1996. Metal analysis and metallography of early metal objects from Denmark. In H. Vandkilde, *From Stone to Bronze: The Metalwork of the Late Neolithic and Earliest Bronze Age in Denmark*. Aarhus: Jutland Archaeological Society, **32**: 321-58.

Ottaway, B. S., 2003. Experimental archaeometallurgy. In Th. Stollner, G. Korlin, G. Steffens, and J. Cierny (eds.), *Man and Mining - Mensch und Bergbau. Studies in honour of Gerd Weisgerber on the Occasion of his 65th Birthday. Der Anschnitt*, **16**: 341-8.

Ottaway, B. S. and Seibel, S., 1997. Dust in the wind: experimental casting of bronze in sand moulds. In *Instrumentum. Monographies 5. Paléométallurgie des Cuivres*. Actes du colloque de Bourg-en-Bresse et Beaune, 17-18 October 1997.

Pigott, V. C., 1999. Reconstructing the copper production process as practised among prehistoric mining/metallurgical communities in the Khao Wong Prachan valley of central Thailand. In S. M. M. Young, A. M. Pollard, P. Budd and R. A. Ixer (eds.), *Metals in Antiquity*. BAR International Series **792**: 10-21. Oxford: BAR Publishing.

Ratka, M., 1996. Numerical simulation of ancient bronze castings. A lecture given at the 13th International Bronze Congress, Cambridge, Massachusetts, May 1996.

Schneider, G., 1989. Investigation of crucibles and moulds from bronze foundries in Olympia and Athens and determination of provenance of bronze statues. In Y. Maniatis (ed.), Archaeometry Proceedings of the 25th International Symposium, 305-20. Oxford: Elsevier.

Schmalfuss, M. and Pernicka, E., (in press). Chemical composition and mechanical properties of Bronze Age ingot torques. In J. P. Northover and J. Toms (eds.) *Founders, Smiths and Platers*. International Conference, Oxford, 1999.

Scott, D., 1991. *Metallography and Microstructure of Ancient and Historic Metals*. Los Angeles: The Getty Conservation Institute.

Shalev, S., 1999. Recasting the Nahal Mishmar hoard: experimental archaeology and metallurgy. *The Beginnings of Metallurgy, Der Anschnitt* , **9**: 295-99.

Srinivasan, S., 1998. Highlights of ancient metallurgy in South India: early use of high-tin bronzes, high-carbon steel, metallic zinc, bronze smelting and solid cast images. In *Proceedings of the Fourth International Conference on the Beginning of the Use of Metals and Alloys, May 25-27, 1998*, Kunibiki Messe, Matsue, Shimane: Japan, 79-84.

Staniaszek, B. E. P. and Northover, J. P., 1983. The properties of leaded bronze alloys. In *Proceedings of the*

22nd Symposium on Archaeometry, University of Bradford, Bradford, UK. March 30th – April 3rd 1982, 262-72.

Stansby, A. W., 1984. The production and finishing of wrought bronze tools and other objects. Unpublished undergraduate dissertation: University of Oxford.

Swiss, A. J. and Ottaway, B. S. (in press). Investment or investment? Casting and using a bronze mould. In J. P. Northover and J. Toms (eds.) *Founders, Smiths and Platers*. International Conference, Oxford, 1999.

Tan Derui and Huang Long, 1996. Shang Zhou Qing Tong Qi Tao Fan Cai Liao Yan Jiu (A study on treatment techniques about ancient clay mould materials), in Ma Chengyuan (ed.), *Wen Wu Bao Hu Ke Xue Lun Wen Ji* (A collection of the Shanghai Museum essays on science of conservation), 47-70. Shanghai: Shanghai Scientific Technological Literature Publishing House.

Tylecote, R.F., 1976. *A History of Metallurgy*. London: The Metals Society.

1986. *The Prehistory of Metallurgy in the British Isles*. London: The Institute of Metals.

1987. *The Early History of Metallurgy in Europe*. London: Longman.

Underwood, L., 1958. Bronze Age technology in western Asia and northern Europe: part 1. *MAN*, **13**: 17 - 23.

Voce, E., 1975. Bronze castings in ancient moulds. In H. Coghlan, *Notes on the Prehistoric Metallurgy of the Copper and Bronze in the Old World*. Occasional Papers on Technology, 4, 2nd edn., 136-9. Oxford: Oxford University Press.

Wang Quanyu, 2002. *Metalworking Technology and Deterioration of Jin Bronzes from the Tianma-Qucun Site, Shanxi, China*. BAR International Series **1023**. Oxford: BAR Publishing.

Wirth, M., 2003. *Rekonstruktion bronzezeitlicher Giessereitechniken mittels numerischer Simulation, giesstechnologischer Experimente und werkstoff-technischer Untersuchungen an Nachguss und Original*. Aachen: Shaker.

www.ingramcontent.com/pod-product-compliance
Lightning Source LLC
LaVergne TN
LVHW072258300926
849160LV00018B/1694
* 9 7 8 1 8 4 1 7 1 6 7 6 3 *